Millets

Millets
Properties, Processing, and Health Benefits

Anil Kumar Siroha
Sneh Punia
Sukhvinder Singh Purewal
Kawaljit Singh Sandhu

CRC Press
Taylor & Francis Group
Boca Raton London New York

CRC Press is an imprint of the
Taylor & Francis Group, an **informa** business

First edition published 2022
by CRC Press
6000 Broken Sound Parkway NW, Suite 300, Boca Raton, FL 33487-2742

and by CRC Press
2 Park Square, Milton Park, Abingdon, Oxon, OX14 4RN

Library of Congress Cataloging-in-Publication Data
Names: Siroha, Anil Kumar, editor.
Title: Millets : properties, processing, and health benefits / Anil Kumar Siroha, Sneh Punia, Sukhvinder Singh Purewal, Kawaljit Singh Sandhu.
Description: First edition. | Boca Raton : CRC Press, 2021. | Includes bibliographical references and index.
Identifiers: LCCN 2021021527 (print) | LCCN 2021021528 (ebook) |
ISBN 9780367562748 (hardback) | ISBN 9781032022581 (paperback) |
ISBN 9781003105947 (ebook)
Subjects: LCSH: Millets. | Millets—Nutrition. | Millets—Utilization.
Classification: LCC SB191.M5 M5555 2021 (print) | LCC SB191.M5 (ebook) |
DDC 633.1/7—dc23
LC record available at https://lccn.loc.gov/2021021527
LC ebook record available at https://lccn.loc.gov/2021021528

ISBN: 978-0-367-56274-8 (hbk)
ISBN: 978-1-032-02258-1 (pbk)
ISBN: 978-1-003-10594-7 (ebk)

Typeset in Palatino
by codeMantra

CONTENTS

PREFACE

Millets are low-cost cereal grains that are widely used in food industries and animal husbandry as an important source of food and feed. Millet grains are rich sources of starch, protein, minerals, vitamins, and specific bioactive compounds with health-benefitting antioxidant properties. Due to their good nutritional and phytochemical profile, they have gained considerable attention as a botanical dietary supplement in many functional foods. Exploring the properties of millets provides a basis for better utilization of millet crops as well as for further development of millet as an industrially important crop. This book will explore knowledge about millets; their production, chemistry, and nutritional aspects; processing technologies; product formulations; etc.

Chapter 1 deals with taxonomy, history, nutritional aspects, and health benefits of millets. Millets are unique among the cereals because of their high calcium, iron, potassium, magnesium, phosphorous, zinc, dietary fiber, polyphenols, and protein content with potential health benefits. Various processing methods to increase the nutritional properties of millets are discussed in Chapter 2. These include milling, decortication, blanching, germination, fermentation, microwave cooking, extrusion cooking, roasting, toasting, and popping. Many anti-nutritional components are reduced after processing, which improves the nutritional profile of millet grains.

Chapter 3 discusses the physical and functional properties of millets that play an important role in the physical behavior and affect the sensory characteristics of foods. Nutritional composition and health benefits of millets are discussed in Chapter 4. Millet grains are good source of dietary as well as health-benefiting components such as carbohydrates, proteins, fiber, minerals/vitamins, and bioactive compounds. Consumption of nutritionally rich millets is beneficial to many diseases including diabetes, celiac disease, cancer, diabetes, and chronic disorders. The major proportion in the grains is occupied by starch followed by protein, fat, fibers, minerals/vitamins, and bioactive compounds. Chapter 5 deals with starch composition, structure, properties, and applications. Starch isolation methods, starch structure, pasting, rheological, morphological, digestibility properties, and applications of millet starches are thoroughly reviewed. Millet starches have comparable properties with other cereal starches such as corn; so, it can be used as a substitute of cereal starches. During processing, the texture and appearance of the product is altered; so, to overcome undesirable changes such as instability of paste under shear, acid, or freezing conditions and poor paste clarity, starch needs to be modified. Different methods of starch modifications (physical, chemical, and enzymatic) are discussed in chapter 6. Food industries have shown their interest in producing more natural food components; thus, there is an increasing need to improve the properties of native starches without using chemical modifications. Chapter 7 deals with the antioxidant profile of millets as well as the different components that are responsible for the antioxidant properties of millets. Millets are superior to rice and wheat and, therefore, provide proteins, minerals, and vitamins where the need for such nutrients is in high demand. Chapter 8 explored various millet-based products, i.e., flat breads, extruded snacks, and various alcoholic and non-alcoholic beverages. Detection of disease-causing pathogens in millets is crucial for the maintenance of sustainability in the farming sector. Chapter 9 focuses on various diseases and postharvest management of millets.

This book is useful for students, academicians, researchers, and other interested professionals working on starch, antioxidants, and new product formulations. This book is designed in such a way that it deals with important aspects related to millets. The authors would appreciate receiving any comments/information for the future course of actions.

AUTHORS

Anil Kumar Siroha, Ph.D., is presently working as assistant professor (C) in the Department of Food Science and Technology, Chaudhary Devi Lal University, Sirsa, Haryana (India). He has both a master's and a doctorate degree in Food Science and Technology. His area of interests include starch, starch modification, and development of new products. He has published more than 20 research papers in national and international journals and several book chapters.

Dr. Siroha edited a book entitled *Pearl Millet: Properties, Functionality and Applications*. He also serves as a reviewer for various national and international journals. He is an active member of Association of Food Scientists and Technologists (AFSTI), Mysore, India.

Sneh Punia, Ph.D., is presently working as assistant professor (C) in the Department of Food Science and Technology, Chaudhary Devi Lal University, Sirsa, Haryana (India). Dr. Punia is involved in the mandated research activities of the institution and has expertise on extraction and functional characterization of antioxidants, starch, their modifications, and functional products.

She has presented her research in various national and international conferences and has published more than 50 research papers/book chapters in national and international journals/books. Dr. Punia has published two edited and one authored book with CRC Press / Taylor & Francis Group. She also serves as the reviewer for various international journals. Dr. Punia received her M.Sc. and doctorate degrees in Food Science and Technology from Chaudhary Devi Lal University.

Sukhvinder Singh Purewal, Ph.D., is presently working as Young Scientist (SYST Scheme), in the Department of Food Science & Technology, Maharaja Ranjit Singh Punjab Technical University, Bathinda, Punjab. His area of interests includes solid-state fermentation, bioactive compounds from natural resources, antioxidants, and food biotechnology. He has published more than 15 research papers in journals of international repute.

He has also worked on a UGC, Delhi-sponsored Major Research Project during 2012–2015. He was awarded for the best paper presentation in national as well as international conferences. He is an active life member of Association of Microbiologists of India (AMI), The Biotech Research Society, India (BRSI), Mycological Society of India (MSI), Association of Food Scientists and Technologists (AFSTI), Mysore, India, and Indian Science Congress Association (ISCA).

Kawaljit Singh Sandhu, Ph.D., is presently working as associate professor and head in the Department of Food Science &Technology, Maharaja Ranjit Singh Punjab Technical University (MRSPTU), Bathinda, Punjab. He has more than 18 years of experience in teaching and research. He joined MRSPTU in 2017 as a founder Head of the Department.

In 2006, he was awarded a Post-Doctoral research fellowship by the Korean Ministry of Education, South Korea, to carry out research on starch structural and digestibility properties. He received the Young Scientist Award from the Association of Food Scientists and Technologists (AFSTI), Mysore, India. Dr. Sandhu has successfully completed one research project funded by UGC, New Delhi. Dr. Sandhu's areas of research interest include characterization, modification, and utilization of starches from different botanical sources, drug delivery, bioactive compounds, and food product development. He has published more than 75 international/national scientific papers and several book chapters.

LIST OF ABBREVIATIONS

ABTS+	(2,2′-Azino-bis(3-ethylbenzothiazoline-6-sulfonic acid))
BC	Before Christ
BMF	Barnyard millet flour
BU	Brabender unit
BV	Breakdown viscosity
CP	Carrot power
Cp	Centi-pascal
CPP	Carrot pomace powder
CTC	Condensed tannin content
CVD	Cardiovascular disease
DNA	Deoxyribonucleic acid
DP	Degree of polymerization
DPPH-2,2	Diphenyl-1-picrylhydrazyl
DS	Degree of substitution
DSF	Soy flour
EPI	Epichlorohydrin
FAO	Food and Agriculture Organization
Fe	Iron
ft	Feet
g	Gram
GF	Gluten-free
GHz	Gigahertz
GI	Glycemic index
g/Kg	Gram per kilogram
G′	Storage modulus
G″	Loss modulus
h	Hour
HHP	High hydrostatic pressure
HMT	Heat moisture treatment
H₃O⁺	Hydroxonium ion
HTST	High-temperature and short-time
HTT	Hydrothermally-treated
K	Consistency index
Kg	Kilogram

lbs	Pounds
m	Meter
MFM	Malted finger millet
Mg	Magnesium
mg	Milligram
mgAAE/g	Milligram ascorbic acid equivalent per gram
mg CE/100 g	Milligram catechin equivalent per 100 gram
MGD	Multi-grain dalia
mg GAE/g	Milligram gallic acid equivalent per gram
MHz	Megahertz
MLP	Moringa leaves power
Mn	Manganese
mPa.s	Milli pascal second
MPP	Mango peel powder
n	Flow behaviors index
OSA	Octenyl succinic anhydride
P	Phosphorus
Pa	Pascal
Pa.s	Pascal second
PMF	Pearl millet flour
$POCl_3$	Phosphoryl chloride
psi	Pound-force per square inch
PT	Pasting temperature
PV	Peak viscosity
RDS	Rapidly digesting starch
RE	Reaction efficiency
RKF	Red kidney bean flour
RS	Resistant starch
RVA	Rapid visco-analyzer
RVU	Rapid visco-analyzer unit
s	second
SDS	Slowly digesting starch
SEM	Scanning electron microscopy
STMP	Sodium trimetaphosphate
STPP	Sodium tripolyphosphate
SV	Setback viscosity
tanδ	Damping factor
T_c	Conclusion temperature
TIA	Trypsin inhibitory activity

T_o	Onset temperature
T_p	Peak temperature
TPC	Total phenolic content
WHO	World Health Organization
XRD	X-ray diffractogram
Zn	Zinc
°C	Degree Celsius
μg/ml	Microgram per milli liter
μmol FAE/g	Micromole ferulic acid equivalents
β	Beta
ΔHg	Enthalpy of gelatinization
σo	Yield stress

1

Millet Grains: Taxonomy, History, and Nutritional Approach

1.1 INTRODUCTION

Shrinking of agricultural land, rapid urbanization, climate change, and tough competition between food and feed industries for existing food and feed crops have limited the available cultivable plant food sources (mainly cereals). In developing tropical countries, these food resources are inadequate to supply proteins for both human and animals. However, the cheapest food materials are those that are derived from plant sources that, although abundant in nature, are still underutilized. At this juncture, identification, evaluation, and introduction of underexploited millet crops, including crops of tribal utility which are generally rich in protein is one of the long-term viable solutions for a sustainable supply of food and feed materials (Nithiyanantham et al., 2019).

Millet is a family of several genera of different gluten-free grains with small, round seeds that are 2–3 mm in diameter, comparable in size and shape to mustard or coriander seeds (Arendt et al., 2008). The term millet is derived from the French word "mille" which means thousand, to suggest that a handful of millet contains up to a thousand grains (Shahidi & Chandrasekara, 2013). Millets are classified with maize, sorghum, and Coix (Job's tears) in the grass subfamily Panicoideae (Yang et al., 2012).

Thus, all millets fall within the grass family (Poaceae or Gramineae) but have two tribes, Paniceae and Chlorideae (Baltensperger & Cai, 2004). The major millet genera, pearl millet (*Pennisetum glaucum*), which comprises 40% of the world production, foxtail millet (*Setaria italica*) (Yang et al., 2012), proso millet or white millet (*Panicum miliaceum*), and finger millet (*Eleusine coracana*) make up the majority of global production. Pearl millet produces the largest seeds and is the variety most commonly used for human consumption (Mariac et al., 2006; ICRISAT, 2007). Minor millets include barnyard millet (*Echinochloa* spp.), kodo millet (*Paspalum scrobiculatum*), little millet (*Panicum sumatrense*), guinea millet (*Brachiaria deflexa*), browntop millet (*Urochloa ramose*), Teff (*Eragrostis tef*), and fonio (*Digitaria exailis*) (ICRISAT, 2007; FAO, 2009; Adekunle, 2012). Millets rank sixth in the world cereal grain production list. In Africa and Asia, these underutilized grains play a major role in the food security of millions of people (Shahidi & Chandrasekara, 2013). Millets are considered as one of the oldest foods, having been cultivated from the time of the early human civilization; a recent archeobotanical study has shown that the common millet was domesticated as a staple food 10,000 years ago in Northern China (Lu et al., 2009). Millets have advantageous characteristics as they are drought- and pest-resistant grains. Besides being rich in minerals and vitamins with low fat, dietary energy, and glycemic index values, millets have been observed to possess numerous documented health benefits.

1.2 TAXONOMY

All millets belong to the order of Poales and to the family of Poaceae (also Gramineae or true grasses). They belong to either of the two subfamilies of Panicoideae or Chloridoideae. The taxonomy of millets is presented in Table 1.1 along with general information about these grains.

1.3 HISTORY

Millets are given lesser preference than cereals because of their strong taste. Cultivation of millets is as old as the beginning of sedentism and civilization in the anthropological history of the world that dates back to around 8000 BC. It is believed that sorghum, finger millet, and pearl millet were of African origin, whereas foxtail millet, proso millet, and

Table 1.1 Taxonomy of Millets

	Pearl millet	Finger millet	Fox millet	Kodo millet	Proso millet	Barnyard millet	Little millet
Scientific name	*Pennisetum glaucum*	*Eleusine coracana* L.	*Setaria italic* L.	*Paspalum scrobiculatum*	*Panicum miliaceum*	*Echinochloa frumentacea*	*Panicum sumatrense*
Other names	Babala, bajra/ bajira, mahangu	Ragi, wimbi	Italian millet, foxtail bristle grass, German millet, Hungarian millet	Koden, kodra	Common millet, broom millet, hog millet, panic millet	Sama/shama, sawa millet, billion dollar grass	Blue panic, sama
Kingdom	Plantae	Plantae	Plantae	Plantae	Plantae	Plantae	Plantae
Subkingdom:	Tracheobionta	Tracheobionta	Tracheobionta	Viridaeplantae	Tracheobionta		
Superdivision	Spermatophyta—seed plants	Spermatophyta—seed plants	Spermatophyta—seed plants	Embryophyta	Spermatophyta	Spermatophyta	
Division	Magnoliophyta—flowering plants	Magnoliophyta—flowering plants	Magnoliophyta—flowering plants	Magnoliophyta—flowering plants	Magnoliophyta—flowering plants		
Class	Liliopsida—monocotyledons	Liliopsida—monocotyledons	Liliopsida—monocotyledons	Liliopsida—monocotyledons	Liliopsida—monocotyledons		Monocotyledonae
Subclass	Commelinidae	Commelinidae	Commelinidae	Commelinidae	Commelinidae	Commelinidae	
Order	Cyperales	Cyperales	Cyperales	Cyperales	Cyperales	Poales	Poales
Family	Poaceae—grass family	Poaceae—grass family	Poaceae—grass family	Poaceae	Poaceae—grass family	Poaceae—grass family	Poaceae—grass family
Genus	Pennisetum-fountain grass	Eleusine	Setaria	Paspalum	Panicum	Echinochloa	Panicum
Species	*Pennisetum glaucum*—pearl millet	*Eleusine coracana*	*Setaria italica*	*Paspalum scrobiculatum*	*Panicum miliaceum* L.	*Echinochloa frumentacea*	*Panicum sumatrense*

3

kodo millet were Asian in origin. Archeological evidence suggests that foxtail millet and proso millet are the oldest of the cultivated millets, even older than rice. During the medieval period, millets had become the principal food of the poor, especially in Europe. This was probably the start of when millets began to be looked upon as poor man's cereal (Kalaisekar et al., 2016).

Pearl millet: Archeobotanical evidence from sub-Saharan West Africa suggested that pearl millet was the predominant, or even the only, cultivated cereal across the region, including in Mauretania, Mali, Ghana, Burkina Faso, and Cameroun (Neumann, 2005; D'Andrea & Casey, 2002; Klee et al., 2004; Fuller et al., 2007). Probably the earliest archeobotanical evidence for domesticated pearl millet cultivation is from Lower Tilemsi Valley in northeastern Mali at ca. 4500 BC (Manning et al., 2011).

Finger millet: Finger millet originated in Ethiopia (Shiihii et al., 2011) before reaching India (Siwela et al., 2010). It is believed that finger millet was diffused from Africa to India through trade during the Bronze Age up to 4000 years ago (Fuller et al., 2011).

Fox millet: It originated in China where it had been cultivated for over 6000 years (Ambigaipalan et al., 2011).

Kodo millet: It is indigenous to India and is believed to have been domesticated some 3000 years ago (House et al., 1995).

Proso millet: Proso millet is believed to have been domesticated ~10,000 years ago. There have been multiple centers of origins proposed such as northwestern China (Bettinger et al., 2007, 2010a,), central China (Lu et al., 2009), and inner Mongolia (Zhao, 2005), with the most recent evidence pointing toward either one domestication event in China or one in China and one in Europe simultaneously (Hunt et al., 2011).

Barnyard millet: Archeological evidence suggests that it was grown in Japan as early as the Yayoi period, dating back some 4–5 millennia (Watanabe, 1970). Another study puts the earliest records of domestication from the Jomon period of Japan in 2000 B.C. (Nesbitt, 2005). Nozawa et al. (2004) showed that *Echinochloa esculenta* was domesticated from a limited part of the *Echinochloa crus-galli* population.

Little millet: This was domesticated in India (de Wet et al., 1983), particularly in the Eastern Ghats of India, where it forms an important part of tribal agriculture. It has historically been grown mainly in India, Myanmar, Nepal, and Sri Lanka (Prasada Rao et al., 1993).

1.4 PRODUCTION

About 80% of millet grains are used for food, while the rest is used as animal fodder and in the brewing industry for alcoholic products (Saleh et al., 2013; Shivran, 2016). Millets are a major source of energy and protein for millions of people in China, Japan, Africa, and India, and especially for people living in hot and dry areas of the world (Amadou et al., 2013). Estimated global cereal production was 28,369,607 tons in 2017 and was forecasted to be 31,019,370 tons in 2018, with the top producing countries being India (11,640,000 tons), Niger (3,856,344 tons), Sudan (2,647,000 tons), Nigeria (2,240,744 tons), Mali (1,840,321 tons), Mainland China (1,565,965 tons), Burkina Faso (1,189,079 tons), Ethiopia (982,958 tons), Chad (756,616 tons), and Senegal (574,000 tons) (FAO, 2018) (Figure 1.1). Asian and African countries are the biggest millet producers. Regionally, Africa occupies the major share (51%), followed by Asia (47%), America (1%), and Europe (1%) for millet production (FAO, 2017) (Figure 1.2). In India, about 9,107,000 hectares of land is under barley cultivation, with a total millet production of 11,640,000 tons (FAO, 2018).

1.5 PLANT DESCRIPTION AND GROWTH AND ENVIRONMENTAL CONDITIONS

Millets exhibit positive agronomic and nutritional characteristics and are well suited for various climates and in crop rotation with other grains while being resistant to certain pests and diseases (Saleh et al., 2013).

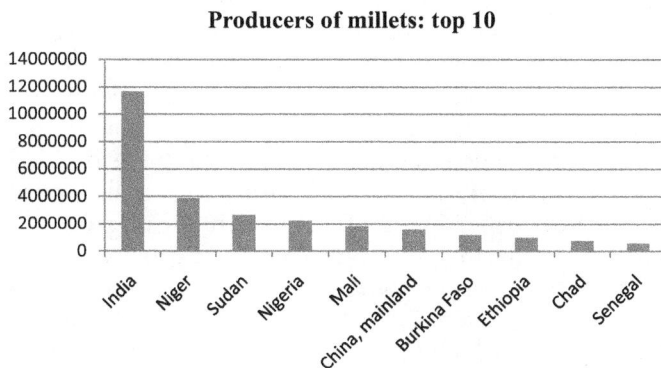

Producers of millets: top 10

Figure 1.1 Top producers of millets worldwide (FAO, 2018).

5

Production share by millets by region

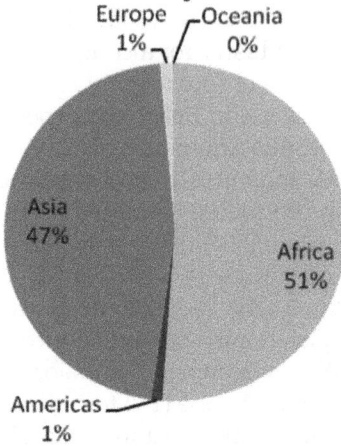

Figure 1.2   Production share of millets worldwide (FAO, 2018).

Therefore, taking into consideration the climate changes that have severely affected the biodiversity, agriculture, and land and water resources, millets might be considered as the "cereal of the future" (Sharma & Gujral, 2019). A brief description of millet plants and grains is presented in Tables 1.2 and 1.3.

Pearl millet (*Pennisetum glaucum*) belongs to the family Poaceae and is widely grown around the world for feed and fodder. Being a drought-tolerant cereal crop, it is grown primarily in India and Africa.

Pearl millet is an upright bunch grass that tillers from the base and has an extensive root system that provides drought tolerance. Stems are 0.5"–1" diameter. It is a leafy plant with leaf blades that are 8–40 inches long and 1/2–3 inches wide. It has a slender stem that is divided into distinct nodes. The leaves of the plant are linear or lance-like, possess small teeth, and can grow up to 1 m (3.3 ft) in length. It can grow with annual precipitation as low as 16–26 in, but it will not grow well above 6500–9000 ft (2000–2700 m) (Hannaway & Larson, 2004). The optimal temperature for growth ranges from 77–86°F (25°C–30°C) (Hannaway & Larson, 2004) to 91–95°F (32°C–35°C) (Newman et al., 2010). It produces approximately 40,000–60,000 seeds/lb (Newman et al., 2010) to 90,900 seeds/lb (Hannaway & Larson, 2004).

Table 1.2 Growth and Environmental Condition of Millets

Millets	Soil Type	Height Range	Temperature	pH	Soil Salinity (dS/m)	Rainfall Required	Maturity Time (Days)	References
Pearl millet	Loamy soils, shallow soils, soils with clay, clay loam and sandy loam texture	Sea level to 2000–2700 m	77–86°F (25°C–30°C)	6.0–7.0 *Can grow up to 8.0 pH	11–12	20–60 cm	60–70	Hannaway and Larson (2004)
Finger millet	Rich loam to poor upland shallow soils	Sea level to 2300 m	26°C–29°C *Lower productivity below 20°C	4.5–7.5	11–12	50–60 cm	90–120	Satish et al. (2016)
Foxtail millet	Sandy to loamy soils	Sea level to 2000 m	Range 5°C–35°C	5.5–7.0	6	30–70 cm	75–90	Brink (2006), Kumar et al. (2018), Hariprassana (2017), Krishnamurthy et al. (2014)
Kodo millet	Fertile to marginal soils	Up to 1500 m	25°C–27°C	–	–	800–1200 mm	100–140	Kannan et al. (2013)

(Continued)

7

Table 1.2 (*Continued*) Growth and Environmental Condition of Millets

Millets	Soil Type	Height Range	Temperature	pH	Soil Salinity (dS/m)	Rainfall Required	Maturity Time (Days)	References
Proso millet	Sandy loam, slightly acidic, saline, low fertility soils	1200–3500 m above sea level	20°C–30°C	5.5–6.5	–	20–50 cm	60–90	Habiyaremye et al. (2017)
Barnyard millet	Medium to heavy soils	Sea level to 2000 m	Range 15°C–33°C Average 27°C–33°C	4.6–7.4	3–5	–	45–70	Farrell (2011), Mitchell (1989)
Little millet	–	Up to 2100 m					80–85	Kannan et al. (2013), DayakarRao et al. (2017)

Table 1.3 Structural Features of Millets

Millets	Type	Shape	Color	Seed Coat	Aleurone Layer
Pearl millet	Caryopsis	Ovoid, hexagonal, globose	Grey, white, yellow, brown, purple	One layered	One layered
Finger millet	Utricle	Globular	White, red, copper brown, violet	Five layered	One layered
Fox millet	Utricle	Oval	Red, black, white, yellow	One layered	One layered
Kodo millet	Utricle	Spherical	Grey brown, brown, dark brown	Multiple layered	One layered
Proso millet	Utricle	Oval	Grey, brown, golden yellow, straw white	One layered	One layered
Barnyard millet	Utricle	Oval	Straw white, dull white	Multiple layered	One layered
Little millet	Utricle	Oval	Grey, straw white	One layered	One layered

Finger millet (*Eleusine coracana* L.), commonly known as ragi, is a millet crop belonging to the family Poaceae of the monocotyledon group (Chandra et al., 2016). Finger millet is the sixth largest crop under cultivation serving as the primary food for rural populations of East and Central Africa and southern India (Vijayakumari et al., 2003). Among the millets, finger millet ranks fourth on the global scale of production after sorghum, pearl millet (*Cenchrus americanus*), and foxtail millet (*Setaria italica*) (Upadhyaya et al., 2007).

Finger millet is an erect, annual grass, growing to about 2–4 ft tall with tillering-tufted stems. When mature, the stems are somewhat laterally flattened, bearing a whorl of 4–6 digitate, straight, or slightly incurved spikes. The spikes are about 1/2 in. broad and 5–6 in. long; the numerous spikelets (about 70) are arranged alternately on the rachis. Each spikelet contains 4–7 seeds varying in diameter from 1 to 2 mm (Reddy et al., 2009). Finger millet grains are globular in shape, and its diameter varies from 1.0 to 1.5 mm (Siwela, 2009; Gull et al., 2014).

Foxtail millet (*Setaria italic* L.), one of the earliest domesticated crops in the world, belongs to the *Setaria* genus of the Poaceae family (Zhang et al., 2017). It is the second largest crop among the millets, cultivated

9

for food in the semi-arid tropics of Asia and as forage in Europe, North America, Australia, and North Africa (Austin, 2006).

An annual grass, foxtail millet forms slender, erect, leafy stems varying in height from 1 to 5 ft. Seeds are borne in a spike-like, compressed panicle resembling yellow foxtail, green foxtail, or giant foxtail. Its small convex seeds are enclosed in colored hulls, with color depending on variety (Baker, 2003).

Kodo millet (*Paspalum scrobiculatum*), subfamily Panicoideae, tribe Paniceae, is a tetraploid ($2n=4x=40$) with a genome size of ca. 1900 Mbp (Jarret et al., 1995). It is also known as cow grass, rice grass, ditch millet, Native Paspalum, or Indian Crown Grass. Kodo millet is an old traditional underutilized crop plant that is used as staple food in Africa (eastern and southern) and Asia (from Near East to Far East).

It is an annual millet that varies in height from 30 to 90 cm and has many basal tillers (between 10 and 48). The inflorescence is small, 2–12 cm. It is self-pollinating, and florets generally remain closed during the flowering period. The grain occurs in a hard husk, making de-branning difficult. The crop is drought-resistant, hardy, and frequently grown in poor soils (Dogget, 1989). It is an extremely hardy and drought-tolerant crop that can survive on marginal soils where other crops may not survive, and it can supply 450–900 kg of grains per hectare (Sharma et al., 2017). Kodo is a monocot, and the seeds are very small and ellipsoidal, being approximately 1.5 mm in width and 2 mm in length; they vary in color from light brown to dark grey (Deshpande et al., 2015).

Proso millet (*Panicum miliaceum* L.) is an annual herbaceous plant in the genus *Panicum*, and it has a chromosome number of $2n=36$, with a basic chromosome number of $x=9$ (Upadhyaya et al., 2016). It is commonly known as broomcorn millet, common millet, hog millet, Russian millet, and so on in different parts of the world. Its popularity is on the rise as it is gluten-free and resistant to high temperatures and drought (Mustač et al., 2020).

Proso millet is a summer annual grass, most frequently grown as a late-seeded summer crop (Baltensperger, 2002; Williams et al., 2007), and can complete its life cycle within 60–100 days (Baltensperger, 2002). Proso millet ranges from 30 to 100 cm tall, with few tillers and an adventitious root system (Baltensperger, 2002). Proso stems and leaves are covered with slight hairs. The leaves may be up to 30 cm long with a short ligule but no auricles. The stem is terminated by a drooping panicle 10–45 cm long that may be open or compact (Vilas et al., 2015). Proso millet can be grown on sandy loam, slightly acidic, saline, and low-fertility soils (Changmei & Dorothy, 2014). As a warm-season crop, proso millet is sensitive to frost and requires warm temperatures for germination and development. Optimal soil temperatures for seed germination range from 20°C to 30°C (Amadou et al., 2013). Grains are round, about 3 mm long and 2 mm wide, and enclosed in a smooth hull, which is typically white or creamy-white, yellow, or red in color, but may be gray, brown, or black (Habiyaremye et al., 2017).

Barnyard millet (*Echinochloa frumentacaea*) is one of the hardiest millets and is called by several names, viz., Japanese barnyard millet, *ooda, oadalu, sawan, sanwa*, and *sanwank*. It is one of the oldest indigenous millet crops in the semi-arid tropics of Africa and Asia and commonly grown in India, Nigeria, Niger, China, Burkina Faso, Mali, Sudan, Uganda, Chad, and Ethiopia (Goswami et al., 2015). Of around 35 species (Table 1.1), two main species, *E. esculenta* (A. Braun) H. Scholz; syn. *E. utilis* Ohwi et Yabuno (Japanese barnyard millet) and *E. frumentacea* Link; syn. *E. colona* var *frumentacea* (Link) Ridl. (Indian barnyard millet), are cultivated (Sood et al., 2015). This millet has a distinct advantage of being a drought-resistant crop and serves as food for people living in these regions. However, it is underutilized in processed food products and commercial food systems (Anis & Sreerama, 2020).

Barnyard millet has a wide adaptation capacity and can grow up to an altitude of 2000 m above mean sea level during summer season (Gupta et al., 2009). Barnyard millet is a tall erect crop that can grow up to 100 cm in height. The inflorescence is 15 cm long, densely branched, and usually has purple-ting with awnless scabrous spikelets. It has a short generation time and the fastest growth among all small millets; it completes its life cycle from seed to seed in 45–60 days and yields about 700–800 kg grain and 1000–2000 kg of straw per hectare (Vilas et al., 2015). The grain is caryopsis and white or yellow in color.

Little millet (*Panicum sumatrense*) is native to India and is called Indian millet. It belongs to the genus *Panicum* and has a chromosome number of $2n=36$, with basic chromosome number of $x=9$. It is mainly cultivated in the Caucasus, China, east Asia, India, and Malaysia. Little millet is adapted to both temperate and tropical climates. It can withstand both drought and water logging (Kalaisekar et al., 2016) (Figure 1.3).

1.6 STRUCTURE OF MILLET GRAINS

Millets have great diversity in grain structure. Millet seeds are of two types, i.e. caryopsis and utricles. In utricles, pericarp surrounds the seed and is attached to the seed at one point. Finger, foxtail, and proso millets

Figure 1.3 A pictorial presentation of millets grains: (a) proso millet; (b) barnyard millet; (c) finger millet; (d) little millet; (e) foxtail millet; (f) pearl millet; (g) kodo millet.

are utricles, wherein the pericarp is loosely attached to the kernel and is removed during harvesting and processing (Taylor & Krugar, 2016). In utricles, the protective layer is testa or seed coat, which is well developed, thick, and strong and provides good protection to the endosperm.

In the caryopsis, the pericarp is fused to the seed. Pearl millet is a type of caryopsis. Sorghum, pearl millet, and teff, the edible grain threshes free from the husk, are naked kernels that can be eaten in their entirety as 'whole grain'.

The principal anatomical components of millets are pericarp, germ or embryo, and endosperm. Pearl millet has a 1–2-layer epicarp, meso-carp, and an endocarp that contains tube and cross cells. Finger millet has a pigmented testa, whereas pearl millet has a thin pigmented or non-pigmented testa/seed coat under the pericarp. The endosperm comprises the major portion of millet grains and consists of an aleurone layer and a peripheral, comeous, and floury area. The aleurone layer completely surrounds the endosperm, and peripheral, comeous, and floury areas are under the aleurone layer. The endosperm contains starch embedded in protein matrix. Fox millet has very little floury endosperm, finger and proso millet are intermediate in texture, and pearl millet has a floury to comeous endosperm. The germ portion may be large relative to the endo-sperm (pearl millet) or very small (proso and finger millets). The struc-tural features of millets are presented in Table 1.3.

1.7 NUTRITIONAL COMPOSITION

Millets have been used as important staples worldwide for centuries. They are unique among the cereals because of their high calcium, iron, potassium, magnesium, phosphorous, zinc, dietary fiber, polyphenols, and protein contents (Devi et al., 2014; Gupta et al., 2014). They are rich in valu-able nutrients such as carbohydrates, proteins, dietary fiber, minerals, and vitamins (Rao, 1986). The protein content of millets is very much comparable to that of other cereals, but they contain lower amounts of carbohydrates. The fat content of common millets, foxtail millets, and barnyard millets is very high and is one of the reasons for the reduction in storage stability. Millets are rich in ash content, showing a higher amount of inorganic matter. Finger millet is known as the richest source of calcium (Chauhan et al., 2018).

Carbohydrates: The average carbohydrates content of millets varies from 56.88 to 72.97 g/100 g. The carbohydrate content of millets consists of starch (60%–75%), non-starchy polysaccharides (15%–20%), and free sug-ars (2%–3%) (Chauhan et al., 2018). The carbohydrate content of barnyard millet is the lowest, and it is the slowest digestible one among all the mil-lets (Veena et al., 2005), which makes it a natural designer food and an ideal food for those with type II diabetics (Ugare et al., 2011). The least carbohydrate content in barnyard millet has also been reported by other researchers (Leder, 2004; Saleh et al., 2013). The amount of total sugars (g/100 g) in pearl millet (2.16–2.78) is higher, followed by that in finger millet (0.59–0.69) and foxtail millet (0.46). Raffinose and stachyose are also higher in pearl millet. Sucrose is the major sugar (g/100 g) in finger millet (0.20–0.24), foxtail millet (0.15), and proso millet (0.66) (Subramanian and Jambunathan, 1980; Subramanian et al., 1981; Murty et al., 1985; Wankhede et al., 1979; Becker & Lorenz, 1978).

Protein: Millets, like all cereals, have low protein content compared to legume and animal sources. Millets abound in vital nutrients, and the protein content of millet grains is considered to be equal or superior in comparison to wheat (*Triticum aestivum*), rice (*Oryza sativa*), maize (*Zea mays*), and sorghum (*Sorghum bicolor*) grains (Kajuna, 2001). Millet proteins are located in the endosperm (80%), germ (16%), and pericarp (3%) (Taylor & Schussler, 1986). All millets have protein content comparable to one another, with an average protein content of 10%–11%, except for finger millet, which has been reported to contain protein in the range of 4.76–11.70 g/100 g (Singh & Raghuvanshi, 2012; Panghal et al., 2006; Baebeau & Hilu, 1993). Ravindran (1991) reported protein contents of 14.4%, 9.8%, and 15.9%, respectively, in common millet, finger millet, and foxtail millet. Pearl and proso millets have approximately 19%–33% higher protein contents than the other millets (Taylor & Kruger, 2016). The protein content in pearl millet ranges from 9% to 21% and is comparable to that of wheat (11.8 g/100 g) but is higher than that in sorghum (10.4 g/100 g), rice (6.8 g/100 g), and maize (4.7 g/100 g) (Kaur et al., 2014).

Saleh et al. (2013) reported that millets are good sources of essential amino acids, except lysine and threonine, but are relatively high in methionine. Finger millet protein is rich in essential amino acids like methionine, valine, and lysine; of the total amino acids present, 44.7% are essential amino acids (Mibithi-Mwikya et al., 2000). Finger millet contains amino acids in concentrations exceeding those of the FAO/WHO-recommended standards. Finger millet protein content is reasonably well balanced; additionally, it also contains more lysine, valine, and threonine than other millets (Ravindran, 1991; Sripriya et al., 1997). Protein fractions of foxtail millet are represented by albumins and globulins (13%), prolamins (39.4%), and glutelins (9.9%). It is thus recommended as an ideal food for diabetics (Saleh et al., 2013; Zhang & Liu, 2015). The protein content in proso millet is around 11% and is richer in essential amino acids (leucine, isoleucine, methionine), compared with wheat (Kalinova & Moudry, 2006).

Lipids: Lipids are relatively minor constitutes in millets. Most of the lipids are located in the scutellar area of the germ. Thus, lipid content is significantly reduced when the germ is removed during decortication or degermination (DayakarRao et al., 2017). The fat is distributed in the bran as well as in the endosperm. The fat content of the millets ranges from 1% to 5%, with lowest in finger and kodo millet (1%) and highest in pearl, foxtail, and proso millets (5%) (Chauhan et al., 2018). Sridhar and Lakshminarayana (1994) compared the lipid contents and compositions of foxtail millet, proso millet, and finger millet grains and found that finger millet contained triacylglycerol, which accounted for 80% of the total lipid, while phospholipid and glycolipid accounted for 14% and 6% of the total lipid content, respectively. The fat generally consists of more than 60% unsaturated fatty acids, including linolenic acid (Ushakumari & Malleshi, 2007).

Fibers: Like other cereals, whole-grain millets contain high amounts of fiber (8%–18%), and high fiber contents of whole-grain millets have also been proposed to contribute to their low glycemic index (GI) (Taylor & Kruger, 2016). The millets with the higher fiber contents provide 55%–66% of the prescribed daily fiber intake, which is associated with health benefits such as reduced risk of cardiovascular disease (CVD), diabetes, and certain cancers. Kodo millet is an excellent source of fiber (9%), as opposed to rice (0.2%) and wheat (1.2%) (Deshpande et al., 2015). Similar high dietary fiber contents of little millet (15.9%) and other millets, viz., foxtail (9.6%), Proso (13.6%), kodo (11.6%), and pearl millets (16.2%), have been documented by Hadimani and Malleshi (1993). Barnyard millet is the richest source of crude fiber (13.6%) (Saleh et al., 2013). Crude fiber content of finger millet is higher than that of rice (0.2%) and wheat (1.2%) (Ravindran, 1991; Sripriya et al., 1997).

Mineral content: The mineral content in millets ranges from 1.7 to 4.3 g/100 g, which is several folds higher than that seen in staple cereals like wheat (1.5%) and rice (0.6%). Finger millet mineral content is markedly higher than that of rice (minerals 0.6%) and wheat (fiber 1.2%) (Ravindran, 1991; Sripriya et al., 1997). It is a rich source of calcium (344 mg/100 g), phosphorous (283 mg), iron (3.9 mg), vitamin B (1.71 mg), vitamin E (22 mg), and other micronutrients (Rathore et al., 2019). Pearl millet is a rich source of iron, zinc, magnesium, copper, manganese, potassium, and phosphorous (Gopalan et al., 2003). Barnyard millet is the richest source of iron (186 mg/kg dry matter) (Saleh et al., 2013), and their consumption can meet the iron requirement of pregnant women suffering from anemia.

15

Foxtail millet contains the highest content of zinc (4.1 mg/100 g) among all other millets and is also a good source of iron (2.7 mg/100 g) (Chandel et al., 2014). These nutrients, i.e. zinc and iron, play an important role in enhancing the immunity.

Vitamin: Millets are also good source of β-carotene and B-vitamins, especially riboflavin, niacin, and folic acid. The thiamine and niacin contents of millets are comparable with that of rice and wheat. The highest thiamine content in millets, i.e. 0.60 mg/100 g, is found in foxtail millet. The riboflavin content of the millets is several-fold higher than that found in staple cereals, and barnyard millet (4.20 mg/100 g) has the highest content of riboflavin, followed by foxtail millet (1.65 mg/100 g) and pearl millet (1.48 mg/100 g) (Kumar et al., 2018). Pearl millet kernels are rich in vitamin A but deficit in vitamins B and C (Gopalan et al., 2003).

Phytochemicals: Millets are a rich source of various phytochemicals including tannins, phenolic acids (free and conjugated forms), anthocyanins, phytosterols, and pinacosanols, and these phytochemicals have potential positive impacts on human health. Millets contain mainly free and conjugated forms of phenolic acids, including derivatives of hydroxybenzoic and hydroxycinnamic acids. In addition, several flavonoids, namely anthocyanidins, flavanols, flavones, flavanones, chalcones, and aminophenolic compounds, are found in millets. Phenolic acids and flavonoids are found in different parts of the grain, and the content and composition vary depending on the type of millet grain (Chandrasekara & Shahidi, 2011). All millet grain fractions possess high antioxidant activity in vitro relative to other cereals and fruits (Awika & Rooney, 2004). It was reported that millets with darkly pigmented testa and pericarp showed a higher phenolic content of soluble phenolic fractions than those with light color testa, such as white or yellow (Chandrasekara & Shahidi, 2010). Pigmented sorghum contains unique anthocyanins that could be potential food colorants (Table 1.4).

1.8 HEALTH BENEFITS

Millets have many nutritional and medical functions (Obilana & Manyasa, 2002; Yang et al., 2012) and are rich in health-promoting phytochemicals; thus, they are considered functional foods (Pathak, 2013). Millets have potential health benefits, and epidemiological studies have shown that consumption of millets reduces the risk of heart disease, protects from diabetes, improves

Table 1.4 Nutritional Compositions of Millets

	Protein (%)	Fat (%)	Ash (%)	Crude fibers (%)	Carbohydrates (%)	Fibers (%)	Minerals									Vitamins		
							Ca (%)	Fe (%)	P (%)	K (%)	Na (%)	Mg (%)	Mn (%)	Zn (%)	A	B1 (mg/100 g)	B2 (mg/100 g)	B3 (mg/100 g)
Pearl millet	9–21	5.6–5.9	1.9–2.3	2	68%	7	0.01	74.90	0.35	0.44	0.01	0.13	18	29.50		0.38	0.22	2.70
Finger millet	9.8%	1–1.7	3	3	72.2%–81.5%	19.1	0.33	46	0.24	0.43	0.02	0.11	7.50	15	6.0	0.2–0.48	0.12	0.30
Fox millet	10.29	3.06	3.10	3	60.9–69.95	19.1	0.01	32.60	0.31	0.27	0.01	0.13	21.90	21.90		0.48	0.12	3.70
Proso millet	11	3.5	3.6	9	56.1–70.4	8.5	0.01	33.10	0.15	0.21	0.01	0.12	18.10	18.10		0.63	0.22	1.32
Kodo millet	8.3	1.4	3.6	9	65	37.8	0.01	7	0.32	0.17	0.01	0.13	–	–		0.32	0.05	0.70
Barnyard millet	6.93	2.02	4.27	2.98	71.87	12.6–13.7	23.16	6.91	–					57.45		0.33	4.20	0.10
Little millet	7.7	4.7			67	12.2	16.06	9.30–20										

Sources: Ramashia (2018), Habiyaremye et al. (2017), IFCT (2017), Verma et al. (2015), Ugare et al. (2014), Kaur et al. (2014), Devi et al. (2014), Saleh et al. (2013), Siwela (2009), Gopalan et al. (2003), Chowdhury and Punia (1997), Hadimani and Malleshi (1993).

the digestive system, lowers the risk of cancer, detoxifies the body, increases immunity in respiratory health, increases energy levels, and improves muscular and neural systems as well as being protective against several degenerative diseases such as metabolic syndrome and Parkinson's disease (Manach et al., 2005; Scalbert et al., 2005; Chandrasekara & Shahidi, 2012). The important nutrients present in millets include resistant starch, oligo-saccharides, lipids, antioxidants such as phenolic acids, avenanthramides, flavonoids, lignans, and phytosterols, which are believed to be responsible for many health benefits (Miller, 2001; Edge et al., 2005). Lee et al. (2010) reported that foxtail and proso millet grain consumption decreased the triacylglycerol level and that foxtail millet specifically reduced C-reactive protein level, an inflammation-related indicator in hyperlipidemic rats, suggesting their potential application in lowering the risk for CVD. Phenolic content as well as composition of different millet grains and avail-able evidences for beneficial health effects emphasize their significance as a functional food ingredient in non-communicable disease reduction and improving health (Shahidi & Chandrasekara, 2013).

1.9 CONCLUSION

Millets are a group of highly variable small-seeded grasses gown world-wide as a source of food; they possess many valuable nutritional compo-nents that enrich the human diet. Millets have been cultivated for centuries due to their versatility, ability to adapt to unfavorable climate and soil conditions, and better characteristics for food and nonfood applications. Millet grains are an excellent source of soluble and insoluble dietary fiber, vitamins, minerals, and phenolic compounds; therefore, the health ben-efits associated with these compounds in millets will stimulate interest among food producers and consumers in using millets for food purposes.

REFERENCES

Adekunle, A. A. 2012. *Agricultural innovation in sub-Saharan Africa: experiences from multiple-stakeholder approaches*. Forum for Agricultural Research in Africa, Ghana. ISBN 978-9988-8373-2-4.

Amadou, I., Gounga, M. E. and Le, G. W. 2013. Millets: nutritional composition, some health benefits and processing–a review. *Emirates Journal of Food and Agriculture* 25: 501–508.

Ambigaipalan, P., Hoover, R., Donner, E., Liu, Q., Jaiswal, S., Chibbar, R. and Seetharaman, K. 2011. Structure of faba bean, black bean and pinto bean starches at different levels of granule organization and their physicochemical properties. *Food Research International* 44(9): 2962–2974.

Anis, M. A. and Sreerama, Y. N. 2020. Inhibition of protein glycoxidation and advanced glycation end-product formation by barnyard millet (*Echinochloa frumentacea*) phenolics. *Food Chemistry* 315: 126265.

Arendt, E. K., Morrissey, A., Moore, M. M. and Dal Bello F. 2008. Gluten-free breads. In: Arendt, E. K., Dal Bello F., editors. *Gluten-Free Cereal Products and Beverages*. San Diego: Academic Press. pp. 289–319.

Austin, D. 2006. Fox-tail millets (*Setaria*: Poaceae): abandoned food in two hemispheres. *Economic Botany* 60: 143–158.

Awika, J. M. and Rooney, L. W. 2004. Sorghum phytochemicals and their potential impact on human health. *Phytochemistry* 65: 1199–1221.

Baebeau, W. E. and Hilu, K. W. 1993. Plant foods. *Human Nutrition* 43(2): 97–104.

Baker, R. D. 2003. Millet production-guide A-414. New Mexico State University, College of Agriculture and Home Economics, Las Cruces, NM.

Baltensperger, D. D. 2002. Progress with proso, pearl and other millets. In: Janick J. and Whipkey A., editors. *Trends in new crops and new uses*. Alexandria: ASHS Press. pp. 100–103.

Baltensperger, D. and Cai, Y. Z. 2004. Millet Minor. In: Wrigley C., Corke H., Walker C. E., editors. *Encyclopedia of Grain Science*, Vol. 2, Amsterdam: Elsevier Science. pp. 261–268.

Becker, R. and Lorenz, K. 1978. Saccharides in proso and foxtail millets. *Journal of Food Science* 43(5): 1412–1414.

Bettinger, R. L., Barton, L. and Morgan, C. 2010a. The origins of food production in north China: a different kind of agricultural revolution. *Evolutionary Anthropology* 19: 9–21.

Bettinger, R. L., Barton, L., Morgan, C., Chen, F., Wang, H. and Guilderson, T. P. 2010b. The transition to agriculture at Dadiwan, People's Republic of China. *Current Anthropology* 51: 703–714.

Bettinger, R. L., Barton, L., Richerson, P. J., Boyd, R., Wang, H. and Choi, W. 2007. The transition to agriculture in northwestern China. *Developments in Quaternary Sciences* 9: 83–101.

Brink, M. 2006. *Setaria italica* (L.) P. Beauv. Record from protabase. PROTA (Plant Resources of Tropical Africa/Ressources végétales de l'Afriquetropicale), Wageningen, Netherlands.

Chandel, G., Kumar, M., Dubey, M. and Kumar, M. 2014. Nutritional properties of minor millets: neglected cereals with potentials to combat malnutrition. *Current Science* 107: 1109–1111.

Chandra, D., Chandra, S. and Sharma, A. K. 2016. Review of Finger millet (*Eleusine coracana* (L.) Gaertn): a power house of health benefiting nutrients. *Food Science and Human Wellness* 5: 149–155.

Chandrasekara, A. and Shahidi, F. 2010. Content of insoluble bound phenolics in millets and their contribution to antioxidant capacity. *Journal of Agriculture and Food Chemistry* 58: 6706–6714.

Chandrasekara, A. and Shahidi, F. 2011. Determination of antioxidant activity in free and hydrolyzed fractions of millet grains and characterization of their phenolic profiles by HPLC–DAD-ESI-MS^n. *Journal of Functional Foods* 3: 144–158.

Chandrasekara, A. and Shahidi, F. 2012. Bioaccessibility and antioxidant potential of millet grain phenolics as affected by simulated in vitro digestion and microbial fermentation. *Journal of Functional Foods* 4: 226–237.

Changmei, S. and Dorothy, J. 2014. Millet-the frugal grain. *International Journal of Scientific Research and Review* 3: 75–90.

Chauhan, M., Sonawane, S. K. and Arya, S. S. 2018. Nutritional and nutraceutical properties of millets: a review. *Clinical Journal of Nutrition and Dietetics* 1(1): 1–10.

Chowdhury, S. and Punia, D. 1997. Nutrient and antinutrient composition of pearl millet grains as affected by milling and baking. *Nahrung* 41: 105–107.

DayakarRao, B., Bhaskarachary, K., Arlene Christina, G. D., Sudha Devi, G., Vilas, A. T. and Tonapi, A. 2017. Nutritional and health benefits of millets. ICAR Indian Institute of Millets Research (IIMR) Rajendranagar, Hyderabad, 112.

D'Andrea, A. C. and Casey, J. 2002. Pearl millet and Kintampo subsistence. *African Archaeological Review* 19: 147–173.

Deshpande, S. S., Mohapatra, D., Tripathi, M. K. and Sadvatha, R. H. 2015. Kodo millet: nutritional value and utilization in Indian foods. *Journal of Grain Processing & Storage* 2: 16–23.

Devi, P. B., Vijayabharathi, R., Sathyabama, S., Malleshi, N. G. and Priyadarisini, V. B. 2014. Health benefits of finger millet (*Eleusine coracana* L.) polyphenols and dietary fiber: a review. *Journal of Food Science & Technology* 51(6): 1021–1040.

de Wet, J. M. J., PrasadaRao, K. E. and Brink, D. E. 1983. Systematics and domestication of *Panicum sumatrense* (Gramineae). *Journal d'agriculture traditionnelle et de botanique appliquée* 30: 159–168.

Dogget, H. 1989. Small millets: a selective over view. In: Seetha ram, A., Riley, K., Harinaryana, G., editors. *Small Millets in Global Agriculture*. New Delhi: Oxford & amp; IBH. pp. 3–18.

Edge, M. S., Jones, J. M. and Marquart, L. 2005. A new life for whole grains. *Journal of American Dietetic Association* 105(12): 1856–1860.

FAOSTAT 2009. *FAO—Food and Agriculture Organization of the United Nations.* Available from FAOSTAT Statistics database-agriculture. Rome, Italy: Food and Agriculture Organization of the United Nations.

FAOSTAT 2017. *FAO—Food and Agriculture Organization of the United Nations.* Available from FAOSTAT Statistics database-agriculture. Rome, Italy: Food and Agriculture Organization of the United Nations.

FAOSTAT 2018. *FAO—Food and Agriculture Organization of the United Nations.* Available from FAOSTAT Statistics database-agriculture. Rome, Italy: Food and Agriculture Organization of the United Nations.

Farrell, W. 2011, Plant guide for billion-dollar grass (Echinochloa frumentacea), USDA-Natural Resources Conservation Service.

Fuller, D. Q., Boivin, N., Hoogervorst, T. and Allaby, R. 2011. Across the Indian Ocean: the prehistoric movement of plants and animals. *Antiquity* 85: 544–558.

Fuller, D. Q., Macdonald, K. and Vernet, R. 2007. Early domesticated pearl millet in DharNema (Mauritania): evidence of crop processing waste as ceramic temper. In: Cappers, R., editor. *Field of change. Proceedings of the 4th international workshop for African Archaeobotany.* Groningen: Barkhuis & Groningen University Library. pp. 71–76.

Gopalan, C., Rama Sastri, B. V. and Balasubramanian, S. C. 2003. *Nutritive value of Indian foods.* Hyderabad: National Institute of Nutrition.

Goswami, D., Gupta, R. K., Mridula, D., Sharma, M. and Tyagi, S. K. 2015. Barnyard millet based muffins: physical, textural and sensory properties. *LWT-Food Science and Technology* 64: 374–380.

Gull, A., Jan, R., Nayik, G. A., Prasad, K. and Kumar, P. 2014. Significance of finger millet in nutrition, health and value added products: a review. *Journal of Environmental Science, Computer Science and Engineering and Technology* 3(3): 1601–1608.

Gupta, A., Mahajan, V., Kumar, M. and Gupta, H. S. 2009. Biodiversity in the barnyard millet (*Echinochloa frumentacea* Link, Poaceae) germplasm in India. *Genetic Resources & Crop Evolution* 56: 883–889.

Gupta, S., Shrivastava, S. K. and Shrivastava, M. 2014. Proximate composition of seeds of hybrid varieties of minor millets. *International Journal of Research in Engineering and Technology* 3: 687–693.

Habiyaremye, C., Matanguihan, J. B., D'AlpoimGuedes, J., Ganjyal, G. M., Whiteman, M. R., Kidwell, K. K. and Murphy, K. M. 2017. Proso millet (*Panicum miliaceum* L.) and its potential for cultivation in the Pacific Northwest, US: a review. *Frontiers in Plant Science* 7: 1961.

Hadimani, N. A. and Malleshi, N. G. 1993. Studies on milling, physico-chemical properties, nutrient composition and dietary fiber content of millets. *Journal of Food Science and Technology* 30(1): 17–20.

Hannaway, D. B. and Larson, C. 2004. *Forage fact sheet: pearl millet (Pennisetum americanum).* Corvallis, OR: Oregon State University.

Hariprassana, K. 2017. Foxtail millet, Setaria italica (L.) P. Beauv. In: Jagananth, P. V., Editor. *Millets and sorghum: biology and genetic improvement.* Wiley: New York. pp. 112–148.

House, L. R., Osmanzai, M., Gomez, M. I., Monyo, E. S. and Gupta, S. C. 1995. Agronomic principles. In: Dendy, D. A. V., editor. *Sorghum and millets: chemistry and technology.* St Paul, MN: American Association for Cereal Chemist. pp. 27–67.

Hunt, H. V., Campana, M. G., Lawes, M. C., Park, Y. J., Bower, M. A., Howe, C. J. and Jones, M. K. 2011. Genetic diversity and phylogeography of broomcorn millet (*Panicum miliaceum* L.) across Eurasia. *Molecular Ecology* 20(22): 4756–4771.

21

ICRISAT 2007. International Crops Research Institute for the Semi-Arid Tropics, annual report. http://test1.icrisat.org/Publications/EBooksOnlinePublications/AnnualReport-2007.pdf.

IFCT, 2017. Longvah, T., et al (editor). Indian food composition tables. Hyderabad: National Institute of Nutrition, Indian Council of Medical Research. http://www.ifct2017.com/frame.php?page=home.

Jarret, R. L., Ozias-Akins, P., Phatak, S., Nadimpalli, R., Duncan, R., et al. 1995. DNA contents in *Paspalum* spp. determined by flow cytometry. *Genetic Resources & Crop Evolution* 42: 237–242.

Kajuna, S. T. A. R. 2001. Millet: post-harvest operations. *Food & Agricultural Organisation* 5: 1–49.

Kalaisekar, A. Padmaja, P. G. Bhagwat, V. R. and Patil, J. V. 2016. *Insect pests of millets - Systematics, Bionomics, and Management*. USA: Academic Press.

Kalinova, J. and Moudry, J. 2006. Content and quality of protein in proso millet (*Panicum miliaceum* L.) varieties. *Plant Foods for Human Nutrition* 61: 45–49.

Kannan, S. M., Thooyavathy, R. A., Kariyapa, R. T., Subramanian, K. and Vijayalakshmi, K. 2013. Seed production techniques for cereals and millets. Seed node of the revitalizing rainfed agriculture network Centre for Indian knowledge systems (CIICS), 1–39.

Kaur, K. D., Jha, A., Sabikhi, L. and Singh, A. K. 2014. Significance of coarse cereals in health and nutrition: a review. *Journal of Food Science & Technology* 51(8): 1429–1441.

Klee, M., Zach, B. and Stika, H. P. 2004. Four thousand years of plant exploitation in the Lake Chad basin (Nigeria), part III: plant impressions in potsherds from the Final Stone Age Gajiganna culture. *Vegetation History & Archaeobotany* 13: 131–142.

Krishnamurthy, L., Upadhyaya, H. D., Gowda, C. L. L., Kashiwagi, J., Purushothaman, R., Singh, S. and Vadez, V. 2014. Large variation for salinity tolerance in the core collection of foxtail millet (*Setaria italica* (L.) P. Beauv.) germplasm. *Crop & Pasture Science* 65(4): 353–361.

Kumar, A., Tomer, V., Kaur, A., Kumar, V. and Gupta, K. 2018. Millets: a solution to agrarian and nutritional challenges. *Agriculture & Food Security* 7(1): 31.

Leder, I. 2004. Sorghum and millets. Cultivated plants, primarily as food sources. In: Gyargy, F., editor. *Encyclopedia of life support systems, UNESCO*. Oxford: Eolss Publishers.

Lee, S. H., Chung, I. M., Cha, Y. S. and Park, Y. 2010. Millet consumption decreased serum concent ration of triglyceride and C-reactive protein but not oxidative status in hyperlipidemic rats. *Nutrition Research* 30: 290–296.

Lu, H., Zhang, J., Liu, K. B., Wu, N., Li, Y., Zhou, K., Ye, M., Zhang, T., Zhang, H., Yang, X. and Shen, L. 2009. Earliest domestication of common millet (*Panicum miliaceum*) in East Asia extended to 10,000 years ago. *Proceedings of the National Academy of Sciences* 106: 7367–7372.

Manach, C., Mazur, A. and Scalbert, A. 2005. Polyphenols and prevention of cardiovascular diseases. *Current Opinion Lipidology* 16: 77–84.

Manning, K., Pelling, R., Higham, T., Schwenniger, J. L. and Fuller, D. Q. 2011. 4500-year old domesticated pearl millet (*Pennisetum glaucum*) from the Tilemsi Valley, Mali: new insights into an alternative cereal domestication pathway. *Journal of Archaeological Science* 38: 312–322.

Mariac, C., Luong, Kapran, I., Mamadou, A., Sagnard, F., Deu, M., Chantereau, J., Gerard, B., Ndjeunga, J., Bezancon, G., Pham, J. and Vigouroux, Y. 2006. Diversity of wild and cultivated pearl millet accessions (*Pennisetum glaucum* [L.] R. Br.) in Niger assessed by microsatellite markers. *Theoretical & Applied Genetics* 114: 49–58.

Mibithi-Mwikya, S., Ooghe, W., Van Camp, J., Nagundi, D. and Huyghebaert, A. 2000. Amino acid profile after sprouting, autoclaving and lactic acid fermentation of finger millet (*Elusine coracana*) and kidney beans (*Phaseolus vulgaris* L.). *Journal of Agricultural & Food Chemistry* 48: 3081–3085.

Miller, G. 2001. Whole grain, fiber and antioxidants. In: Spiller, G. A., editor. *Handbook of dietary fiber in Human Nutrition*. Boca Raton, FL: CRC Press. pp. 453–460.

Mitchell, W. A. 1989. Japanese millet ("Echinochloa Crusgalli" var. "frumentacea"): Section 7.1. 6, US Army Corps of Engineers wildlife resources management manual.

Murty, D. S., Singh, U., Suryaprakash, S. and Nicodemus, K. D. 1985. Soluble sugars in five endosperm types of sorghum. *Cereal Chemistry* 62(2): 150–152.

Mustač, N. C., Novotni, D., Habuš, M., Drakula, S., Nanjara, L., Voučko, B. and Ćurić, D. 2020. Storage stability, micronisation, and application of nutrient-dense fraction of proso millet bran in gluten-free bread. *Journal of Cereal Science* 91: 102864.

Nesbitt, M. 2005. Grains. In: Prance, S. G. and Nesbitt, M., editors. *The cultural history of plants*, 57/435. New York: Routledge Publishers.

Neumann, K. 2005. The romance of farming: plant cultivation and domestication in Africa. In: Stahl, A. B., editor. *African archaeology. A critical introduction*. Oxford: Blackwell Publishing Ltd. pp. 249–275.

Newman, Y., Jennings, E. D., Vendramini, J. and Blount, A. 2010. Pearl millet (Pennisetum glaucum): overview and management. *SSAGR-337, one of a series of the Agronomy Department, Florida Cooperative Extension Service*, Institute of Food and Agricultural Sciences, University of Florida.

Nithiyanantham, S., Kalaiselvi, P., Mahomoodally, M. F., Zengin, G., Abirami, A. and Srinivasan, G. 2019. Nutritional and functional roles of millets–a review. *Journal of Food Biochemistry* 43(7): e12859.

Nozawa, S., Nakai, H. and Sato, Y. 2004. Characterization of microsatellite and ISSR polymorphisms among *Echinochloa* (L.) Beauv. spp. in Japan. *Breeding Research (Japan)* 6: 87–93.

Obilana, A. B. and Manyasa, E. 2002. Millets. In Belton, P. S. and Taylor, J. R. N., editors. *Pseudocereals and Less Common Cereals*. Berlin: Springer. pp. 177–217.

Panghal, A., Khatkar, B. S. and Singh, U. 2006. Cereal proteins and their role in food industry. *Indian Food Industry* 25(5): 58–62.

Pathak, H. C. 2013. *Role of millets in nutritional security of India*. New Delhi: National Academy of Agricultural Sciences. pp. 1–16.

PrasadaRao, K., De Wet, J., Gopal Reddy, V. and Mengesha, M. 1993. Diversity in the small millets collection at ICRISAT. In: Riley, K. W., Gupta, S. C., Seetharam, A. and Mushonga, J. N., editors, *Advances in small millets*. New Delhi: Oxford and IBH Publishing Co. Pvt. Ltd.

Ramashia, S. E. 2018. Physical, functional and nutritional properties of flours from finger millet (*Eleusine coracana*) varieties fortified with vitamin B and zinc oxide (Doctoral dissertation).

Rao, M. A. 1986. Rheological properties of fluid foods. In: Rao, M. A. and Rizvi, S. S. H., editors. *Engineering properties of foods*. New York: Marcel Dekker INC. pp. 121–177.

Rathore, T., Singh, R., Kamble, D. B., Upadhyay, A. and Thangalakshmi, S. 2019. Review on finger millet: processing and value addition. *The Pharma Innovation Journal* 8(4): 283–291.

Ravindran, G. 1991. Studies on millets: proximate composition, mineral composition, and phytate and oxalate contents. *Food Chemistry* 39: 99–107.

Reddy, V. D., Rao, K. V., Reddy, T. P. and Kishor, P. B. K. 2009. Finger millet. *Compendium of Transgenic Crop Plants*: 191–198.

Saleh, A. S. M., Zhang, Q., Chen, J. and Shen, Q. 2013. Millet grains: nutritional quality, processing, and potential health benefits. *Comprehensive Reviews in Food Science and Food Safety* 12: 281–295.

Satish, L., Rathinapriya, P., Rency, A. S., Ceasar, S. A., Prathibha, M., Pandian, S. and Ramesh, M. 2016. Effect of salinity stress on finger millet (*Eleusine coracana* (L.) Gaertn): histochemical and morphological analysis of coleoptile and coleorhizae. *Flora-Morphology, Distribution, Functional Ecology of Plants* 222: 111–120.

Scalbert, A., Manach, C., Morand, C., Remesy, C. and Jimenez, L. 2005. Dietary polyphenols and the prevention of diseases. *Critical Reviews in Food Science and Nutrition* 45: 287–306.

Shahidi, F. and Chandrasekara, A. 2013. Millet grain phenolics and their role in diseases risk reduction and health promotion – review. *Journal of Functional Foods* 5(2): 570–581.

Sharma, B. and Gujral, H. S. 2019. Modulation in quality attributes of dough and starch digestibility of unleavened flat bread on replacing wheat flour with different minor millet flours. *International Journal of Biological Macromolecules* 141: 117–124.

Sharma, S., Sharma, N., Handa, S. and Pathania, S. 2017. Evaluation of health potential of nutritionally enriched Kodo millet (*Eleusine coracana*) grown in Himachal Pradesh, India. *Food Chemistry* 214: 162–168.

Shiihii, S. U., Musa, H., Bhati, P. G. and Martins, E. 2011. Evaluation of physicochemical properties of Eleusine coracana starch. *Nigerian Journal of Pharmaceutical Sciences* 10(1): 91–102.

Shivran, A. C. 2016. Biofortification for nutrient-rich millets. In: Singh, U., Praharaj, C. S., Singh and Singh, S. S., editors. *Biofortification of food crops*. New Delhi: Springer. pp. 409–420.

Singh, P. and Raghuvanshi, R. S. 2012. Finger millet for food and nutritional security. *African Journal of Food Science* 6(4): 77–84.

Siwela, M. 2009. Finger millet grain phenolics and their impact on malt and cookie quality (Doctoral dissertation, University of Pretoria).

Siwela, M., Taylor, J. N. R., de Milliano, W. A. J. and Doudu, K. G. 2010. Influence of phenolics in finger millet on grains and malt fungal load, and malt quality. *Food Chemistry* 121: 443–449.

Sood, S., Khulbe, R. K., Gupta, A. K., Agrawal, P. K., Upadhyaya, H. D. and Bhatt, J. C. 2015. Barnyard millet–a potential food and feed crop of future. *Plant Breeding* 134(2): 135–147.

Sridhar, R. and Lakshminarayana, G. 1994. Contents of total lipids and lipid classes and composition of fatty acids in small millets: foxtail (*Setaria italica*), proso (*Panicum miliaceum*), and finger (*Eleusine coracana*). *Cereal Chemistry* 71: 355–358.

Sripriya, G., Antony, U. and Chandra, T. S. 1997. Changes in carbohydrate, free amino acids, organic acids, phytate and HCl extractability of minerals during germination and fermentation of finger millet (*Eleusine coracana*). *Food Chemistry* 58(4): 345–350.

Subramanian, V. and Jambunathan, R. 1980.Traditional methods of processing sorghum (*Sorghum bicolor* L. *Moench*) and pearl millet (*Pennisetum americanum* L.) grains in India. *International Association for Cereal Chemistry* 10: 115–118.

Subramanian, V., Jambunathan, R. and Suryaprakash, S. 1981. Sugars of pearl millet [*Pennisetum americanum* (L.) Leeke] grains. *Journal of Food Science* 46(5): 1614–1615.

Taylor, J. R. N. and Kruger, J. 2016. Millets. *Encyclopedia of Food and Health*: 748–757.

Taylor, J. R. N. and Schussler, L. 1986. The protein composition of the different anatomical parts of sorghum grain. *Journal of Cereal Science* 4: 361–369.

Ugare, R., Chimmad, B., Naik, R., Bharathi, P. and Itagi, S. 2011. Glycemic index and significance of barnyard millet (*Echinochloa frumentacae*) in type II diabetics. *Journal of Food Science and Technology* 51: 392–395.

Ugare, R., Chimmad, B., Naik, R., Bharati, P. and Itagi, S. 2014. Glycemic index and significance of barnyard millet (*Echinochloa frumentacae*) in type II diabetics. *Journal of Food Science and Technology* 51(2): 392–395.

Upadhyaya, H., Gowda, C. and Reddy, V. G. 2007. Morphological diversity in finger millet germplasm introduced from Southern and Eastern Africa. *Journal of Semi-Arid Tropical Agricultural Research* 3: 1–3.

Upadhyaya, H. D., Vetriventhan, M., Dwivedi, S. L., Pattanashetti, S. K. and Singh, S. K. 2016. Singh, M. and Upadhyaya, H. D. (Editor). Proso, barnyard, little, and kodo millets. In: *Genetic and genomic resources for grain cereals improvement*. UK: Academic Press, UK. pp. 321–343.

Ushakumari, S. R. and Malleshi, N. G. 2007. Small millets: nutritional and technological advantages. Food uses of small millets and avenues for further processing and value addition, Krishnegowda, K. and Seetharam, editors. All India Coordinated Small Millets Improvement Project, ICAR, UAS, Bangalore.

Veena, S., Chimmad, B. V., Naik, R. K. and Shanthakumar, G. 2005. Physico-Chemical and Nutritional Studies In Barnyard Millet. Karnataka. *Journal of Agricultural Sciences* 18: 101–105.

Verma, S., Srivastava, S. and Tiwari, N. 2015. Comparative study on nutritional and sensory quality of barnyard and foxtail millet food products with traditional rice products. *Journal of Food Science & Technology* 52(8): 5147–5155.

Vijayakumari, J., Mushtari Begum, J., Begum, S. and Gokavi, S. 2003. Sensory attributes of ethnic foods from finger millet (*Eleusine coracana*). Proceeding of the National Seminar on Processing and Utilization of Millet for Nutrition Security: Recent Trends in Millet Processing and Utilization. Hisar: CCSHAV. pp. 7–12.

Vilas, A. T., Bhat, B. V. and Kannababu, N. 2015. Millet seed technology. https://www.millets.res.in/books/Part-2.pdf.

Wankhede, D. B., Shehnaj, A. and Rao, M. R. 1979. Carbohydrate composition of finger millet (*Eleusinecoracana*) and foxtail millet (*Setaria italica*). *QualitasPlantarum* 28(4): 293–303.

Watanabe, N. 1970. A spodographic analysis of millet from prehistoric Japan. *Journal of the Faculty of Science, University of Tokyo* 3: 357–379.

Williams, M. M., Boydston, R. A. and Davis, A. S. 2007. Wild proso millet (*Panicum miliaceum*) suppressive ability among three sweet corn hybrids. *Weed Science* 55: 245–251.

Yang, X., Wan, Z., Perry, L., Lu, H., Wang, Q., Zhao, C. and Wang, T. 2012. Early millet use in northern China. *Proceedings of the National Academy of Sciences* 109(10): 3726–3730.

Zhang, L., Li, J., Han, F., Ding, Z. and Fan, L. 2017. Effects of different processing methods on the antioxidant activity of 6 cultivars of foxtail millet. *Journal of Food Quality*: 1–9.

Zhang, L. Z. and Liu, R. H. 2015. Phenolic and carotenoid profiles and antiproliferative activity of foxtail millet. *Food Chemistry* 174: 495–501.

Zhao, Z. 2005. Palaeoethnobotany and its new achievements in China. *Kaogu (Archaeology)* 7: e49.

2

Effect of Processing on Millet Properties

2.1 INTRODUCTION

Millets are considered as food security crops due to their sustainability in adverse agro-climatic conditions (Ushakumari et al., 2004), and they have potential to broaden the genetic diversity in the food basket and improve food and nutrition security (Mal et al., 2010). Millet grains, before consumption and for preparing food, are usually processed (Adebiyi et al., 2018) to improve their edible, nutritional, and sensorial properties by using traditional processing techniques (Nazni & Devi, 2016). Physical (milling, decortication, cooking, roasting, blanching, extrusion, and popping), chemical, and biological processes (fermentation and germination) are used to prepare millets for food (Dias-Martins et al., 2018). But negative changes in the nature of millets are unavoidable because the industrial method of processing is not well developed compared to that for other cereals. The processing techniques aim to increase the physico-chemical accessibility of micronutrients, decrease the content of antinutrients, or increase the content of compounds that improve bioavailability (Hotz & Gibson, 2007). This chapter aims to understand the influence of different processing methods (dehulling, pearling, milling, germination, fermentation, and thermal treatments like roasting, extrusion, cooking, etc.) on the nutritional and antioxidant properties of millets (Tables 2.1–2.4 and Figure 2.2).

Table 2.1 Importance of Millet Processing

Factors	Importance	References
Improve nutritional quality	Processing increases the protein, total dietary fiber, and total phenolic content of millet grain.	Sharma et al. (2018a)
Shelf life	Inactivates lipase activity.	Yadav et al. (2012)
Organoleptic properties	Improves sensorial, nutritive value, and nutrient bioavailability.	Liu et al. (2012)
Antinutritional factor	Soaking and cooking improves antinutrient content and bioavailability of minerals.	Lestienne et al. (2005b); Subhash et al. (2015)
Digestibility	Fermentation improves digestibility of grains.	Ali et al. (2003); Hole et al. (2012)
Bioactive properties	Germination process increases the total antioxidant activity, reducing power, DPPH activity, and hydrogen peroxide scavenging activities of foxtail millet.	Sharma et al. (2018b)

2.2 EFFECT OF PROCESSING ON NUTRITIONAL QUALITY

To enhance millet nutritional value and utilization, various processing methods are used. Related to improvement of nutritional characteristics, sensory properties, and convenience, there are some processing technologies that are used in the manufacturing of food products (Saleh et al., 2013).

2.3 MILLING

Milling generally involves removal of bran, i.e., the pericarp, the seed coat, the nucellar epidermis, and the aleurone layer. Milling of small millets is done by adoption of both wheat and rice-milling techniques (Kulkarni et al., 2018). The milling and polishing of millet grains was also attempted with other processing equipment or machines that were specifically not designed for millets for a long period (Jaybhaye et al., 2014). The successful effort in this direction is made by Singh (2010) who tried to mechanize the milling process by developing a dehuller for barnyard millet and optimized the machine

Table 2.2 Various Methods of Processing

Methods	Definition	References
Fermentation	A process in which plant and animal tissues are subjected to the action of microbes and enzymes to give desirable changes and to modify food quality.	Sandhu et al. (2017)
Cooking	It is a type of hydrothermal treatment that induces significant changes in the chemical composition, affecting the bioaccessibility and the concentration of nutrients and health-promoting compounds.	Pellegrini et al. (2010)
Toasting	Toasting is a rapid processing method that uses dry heat treatment for a short time.	Siroha and Sandhu (2017)
Malting	Malting is a process that involves steeping, germination under controlled conditions, and kilning.	Chavan et al. (1989)
Extrusion cooking	A rapid processing method involves high temperature and cooking pressure and short time; this is used to prepare a variety of processed foods.	Sharma et al. (2012)
Soaking	Soaking is a domestic technological treatment that involves hydration of seeds in water for a few hours to allow the seeds to absorb water.	Embaby (2010)
Milling	An intermediate step in the post-production of grains, defined as the act or process of grinding the grain into flour.	Bender (2006)
Popping	It is a procedure of dry heat application on kernels of seeds until internal moisture expands, for preparation of healthy ready-to-eat snacks.	Chauhan and Sarita (2018)
Microwave treatment	It is an electro-heat technique that converts electrical energy into thermal energy on a frequency range of 300 MHz to 300 GHz, and food applications are in the range of 915 MHz to 2.45 GHz.	Gavahian et al. (2019)

parameters for maximizing efficiency and minimizing specific energy consumption and broken grains. Due to the small grain size and unique grain morphology of millets when compared with other cereals, decortication by known cereal-milling methodologies has been difficult, resulting in lower yields and higher kernel breakage (Shobana & Malleshi, 2007). The parboiling method showed significant increase in decortication yield and decrease in kernel breakage for millets (Young et al., 1990).

Table 2.3 Effect of Different Treatments on Nutritional Quality of Millet

Methods	Effects on Nutritional Quality	References
Decortication	Decortication is removal of pericarp and germ grain that improves nutrients bioavailability, storage stability, and sensorial characteristics of millets.	Bora et al. (2019)
Puffing or popping	Enhances color, appearance, taste, and aroma and improves the nutritional profile of foods.	Verma and Patel (2013), Saleh et al. (2013)
Extrusion	Production of precooked and dehydrated foods. Helps solve the problem of malnutrition.	Divate et al. (2015), Rathore et al. (2016)
Roasting	Roasting increases the energy value, carbohydrate, ash, and iron content of the flours.	Obadina et al. (2016)
Fermentation	SSF is generally employed to enhance the nutraceutical properties of pearl millet.	Salar and Purewal (2016)
Soaking	Processing (dehulling, soaking, and cooking) decreases the antinutrients and improves the mineral bioavailability (iron and zinc) and in vitro protein digestibility.	Pawar and Machewad (2006)
Germination	Germination of millet grains increases the dietary fibers, protein, and total phenolic content.	Sharma et al. (2018a)

Generally, finger millet is pulverized to flour for preparation of food products. First, it is cleaned to remove foreign materials such as stones, chaffs, and stalks, then passed through abrasive or friction mills to separate out glumes (non-edible cellulosic tissue), and then pulverized. Normally, it is pulverized in a stone mill or iron disc or emery-coated disc mills (Gull et al., 2016). Decortication, milling, and sieving of millet grains are mostly carried out manually; therefore, there is a need for convenient and motorized milling technology for millet grains to provide a large amount of flour to ensure a consistent source for industrial food uses at commercial scale to help in promoting their utilization

Table 2.4 Effect of Various Processing Methods on Antioxidant Properties of Millets

Processing	Processing Methods	Effect of Processing	References
Cooking	Grains were cooked in 150 ml boiling distilled water on a hot plate.	AOA, TPC, TFC, and ABTS+ activities were decreased, while MCA was increased.	Siroha and Sandhu (2017)
	Boiling of millet grains in distilled water (300 ml) on a hot plate for 15 min with stirring using a glass rod.	TPC, DPPH, and hydroxyl radicals scavenging activity and hydrogen peroxide scavenging activity were decreased, while the reverse was observed for oxygen radical absorption capacity.	Chandrasekara et al. (2012)
Toasting	Pearl millet grains were conditioned to a 10% moisture content for maintaining uniformity during toasting at 115°C in an oven.	AOA, TPC, TFC, ABTS+, and MCA were increased.	Siroha and Sandhu (2017)
Dry heat treatment	Whole grains were heated in hot oven at 110°C for an hour.	Decrease in total phenols was observed.	Nithya et al. (2007)
Dehulling	Millet grains were dehulled using a seed buro hand grinder.	TPC and hydroxyl radical scavenging activity were decreased, whereas increase was observed for H_2O_2 scavenging activity.	Chandrasekara et al. (2012)
	Mechanical dehuller.	Reduction in TPC was observed.	Mohamed et al. (2010)

(Continued)

31

Table 2.4 (*Continued*) Effect of Various Processing Methods on Antioxidant Properties of Millets

Processing	Processing Methods	Effect of Processing	References
Germination	Seeping of seeds in water followed by keeping in folded filter paper with continuous moisture for 48 h, allowing the seeds to geminate.	Decrease in total phenols was observed.	Nithya et al. (2007)
Soaking	Soaking of seeds in water for 14 h.	Decrease in total phenols was observed.	Nithya et al. (2007)
Fermentation	Solid-state fermentation of grains was done with *Aspergillus oryzae*.	DPPH (% inhibition), ABTS$^+$ (% inhibition), FRAP value, and reducing power were increased with increase in incubation period.	Salar and Purewal (2016)

(Saleh et al., 2013). Total carbohydrate content increased in milled grain as a result of the debranning process. The calcium and phosphorus contents of brown grains were 13.1 and 412.4 mg/100 g, respectively. Milling removes nearly 66% of calcium and about 36% of phosphorus from brown grains (Hadimani & Malleshi, 1993). The milling method of millets is described in Figure 2.1.

2.4 DECORTICATION/DEHULLING

Millets were earlier decorticated at the household level by hand pounding. Nowadays, these are milled in rice-milling machines, with slight modification of the process. The decortication process, also known as dehulling, entails removal of the bran (pericarp & germ) of the millet grains, in order to promote improvements in the quality attributes, such as palatability, grain coloration, and reduction of phytic acid and fat content (Hama et al., 2011). The germ of millets, owing to its high lipid content, can be one of the factors leading to rancidity development; this, in turn, can lead to a reduced shelf life and loss of sensory properties (Nantanga et al., 2008).

```
┌─────────────────────────────┐
│           Millets           │
└─────────────────────────────┘
              ⇩
┌─────────────────────────────┐
│     Cleaning of millets     │
└─────────────────────────────┘
              ⇩
┌─────────────────────────────┐
│          Dehulling          │──────────────┐
└─────────────────────────────┘              │
              ⇩                              │
┌─────────────────────────────┐              │
│          Aspiration         │              │
└─────────────────────────────┘              │
              ⇩                              │     Un-hulled grains
┌─────────────────────────────┐              │
│    Dehulled grains with     │              ▲
│   some un-hulled grains     │              │
└─────────────────────────────┘              │
              ⇩                              │
┌─────────────────────────────┐              │
│     Gravity separation      │──────────────┘
└─────────────────────────────┘
              ⇩
┌─────────────────────────────┐
│       Dehulled grains       │
└─────────────────────────────┘
              ⇩
┌─────────────────────────────┐
│    Debranning/Polishing     │
└─────────────────────────────┘
              ⇩
┌─────────────────────────────┐
│    Pulverising in Hammer    │
│      mill/roller mill       │
└─────────────────────────────┘
              ⇩
┌─────────────────────────────┐
│            Flour            │
└─────────────────────────────┘
```

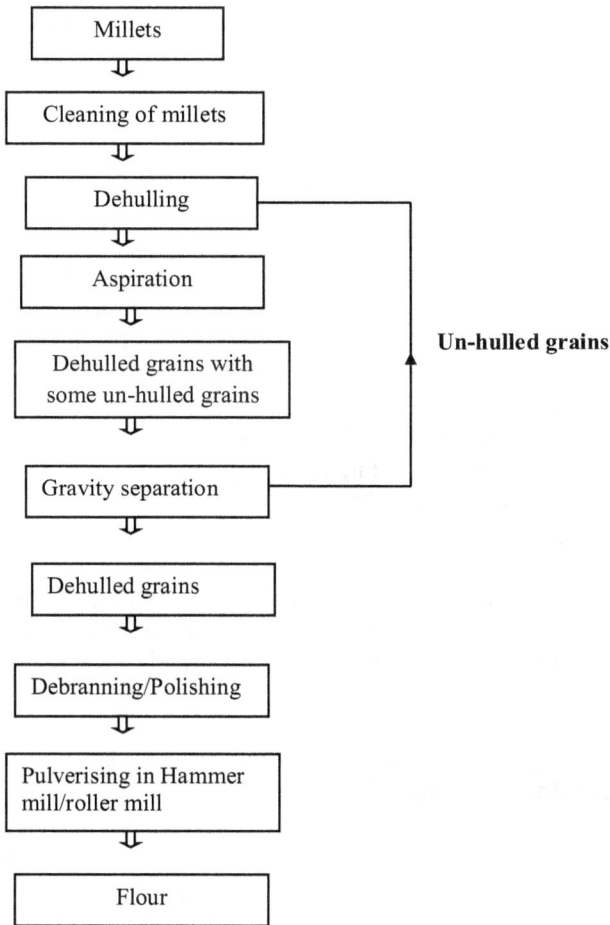

Figure 2.1   Milling of millets.

Bora et al. (2019) stated that the removal of pericarp and germ from the endosperm could improve bioavailability of nutrients, storage stability, and the sensory properties of millets.

Dias-Martins et al. (2019) studied decortication of pearl millet and found that decortications reduced the number of 2.38 mm diameter grains by 46% and increased the number of 1.68 mm diameter grains by 28%. The extraction rate was 90.9%, and around 2.8% of the grains were broken due to the mechanical decortications method used. Decortication can be used to obtained light-colored grains with better palatability, taste, and texture. It promotes changes in grain color by removing the bran, which in turn reduces flavonoids content up to 50% (Dias-Martins et al., 2018). Centrifugal shellers can be used to dehull/decorticate the small millets (Jaybhaye et al., 2014). Abrasive mills (Ayo & Olawale, 2003) or disks with mechanical dehullers are still used for decortication purposes. About 12%–30% of the outer grain surface is removed by decortication; decortication beyond this limit causes substantial loss of ash, fat, micronutrients, fiber, proteins, and amino acids such as lysine, histidine, and arginine (Rai et al., 2008). Serna-Saldivar et al. (1994) reported that although decortication improved protein and dry matter digestibility, due to the removal of the pericarp and germ, reduction in protein, fat, insoluble dietary fiber, ash, lysine, and tryptophan was also observed. Shobana and Malleshi (2007) reported that protein, fat, calcium, and phosphorus contents of the decorticated millet were 6.3%, 0.9%, 0.18%, and 0.10% and were lower by 22%, 40%, 43%, and 48%, respectively, than those found in the native millet. The reduction in some of the nutrients could be mainly due to separation of the seed coat, as it has been reported that the seed coat contains about 28% protein, 49% calcium, and 14% phosphorus (Kurien et al., 1959). Chandrasekara et al. (2012) reported that the loss of TPC in dehulled grain compared with the corresponding whole grain is 78%, 21%, 12%, 65%, 72%, 35%, and 2% for kodo, finger (Ravi), finger (local), foxtail, proso, little, and pearl millets, respectively. Hag et al. (2002) also showed that dehulling decreased the TPC of pearl millets.

2.5 FERMENTATION

Fermentation is a metabolic process that converts complex material into simpler forms with the help of microorganisms. It is the most effective and oldest method of processing and preserving foods. Salar and Purewal (2016) stated that solid-state fermentation could be successfully used to further develop the nutraceutical properties of pearl millets. Among the treatment methods, grain fermentation effectively decreased the inhibitory factors but simultaneously induced organoleptic changes, affecting sensory characteristics (Lestienne et al., 2005b). Fermentation leads to an altered ratio of nutritive and antinutritive components of plants, which affects product properties such as bioactivity and digestibility

(Heinio et al., 2003; Katina et al., 2007). Various microbial strains have been used for the fermentation of foxtail millet flour, such as *Lactobacillus paracasei* Fn032 (Amadou et al., 2014, 2013a, b), *Lactobacillus acidophilus*, *Lactobacillus rhamnosus*, *Bifidobacterium bifidus*, *Bifidobacterium longum* (Farooq et al., 2013), *Saccharomyces boulardii*, and *Lactobacillus acidophilus* (Pampangouda et al., 2015). During fermentation, the grain constituents are modified by the action of both endogenous and bacterial enzymes, thereby affecting their structure, bioactivity, and bioavailability (Hole et al., 2012). Davidek et al. (1990) have reported changes during fermentation including destruction of inhibitors, breakdown of protein, and increase in amino nitrogen. Simply defined, it is a process performed to transform substrates into new products through the action of microorganisms. Biochemical changes that occur during this process lead to the modification of the substrate and production of volatiles (Singh et al., 2015). The fermentation process leads to the activation of enzymes, adjustment of pH, and performance enhancement of certain enzymes, including proteases, amylases, and hemicellulases (Jay et al., 2005). Adebiyi et al. (2016) observed that the higher crude protein values of fermented flour, malted flour, fermented biscuit, and malted biscuit samples are due to the accumulation of proteins and production of some additional amino acids in the samples as a result of fermentation and malting. These observations are also similar to the reports of increase in protein contents of millet due to fermentation and malting (Inyang & Zakari, 2008). Ali et al. (2003) stated that fermentation is one of the processes that decrease the levels of antinutrients in food grains and increase the protein availability, in vitro protein digestibility, and nutritive value. Elyas et al. (2002) observed that fermentation for 36 h at room temperature is found to cause no change in the tannin content of fermented dough for the two millet cultivars. This result agrees with that of Agte et al. (1997) who reported that the levels of tannins in pearl millet were unaffected by fermentation. Many studies report an increase in the solubility of iron and zinc after fermentation, but solubility may not be a reliable indicator of bioavailability as it does not take into consideration the molecular size of the compounds, which is critical for absorption (Etcheverry et al., 2012). Fermentation is also found to decrease trypsin inhibitory activity (TIA), amylase inhibitor activity, phytic acid, and tannins (Osman, 2004; Ejigui et al., 2005; Eltayeb et al., 2007; Abdel-Haleem et al., 2008). Osman (2011) identified that the fermentation process was found to cause no changes in protein, lipid, and ash contents, but it significantly reduced carbohydrate content. The initial drop in carbohydrate content was attributed to the action of microbial α- and β-amylases, whereas the increase at 16 and 20 h could be due to the termination of starch degradation by low pH, which inhibits amylase activity (El-Tinay et al., 1979) and/or to the presence of tannins that inhibit amylotic enzymes. Use of endogenous grain microflora in finger millet flour during fermentation shows a significant reduction in the amount of antinutrient factors such as trypsin inhibitor activity (TIA) by 32%, tannins by 52%, and phytates by 20% (Antony & Chandra, 1998). Singh and Raghuvanshi (2012) have reported that fermentation also increases mineral availability and accessibility, such as calcium by 20%, zinc by 26%, phosphorous by 26%, and iron by 27%. With these desirable benefits, fermentation has been considered as an effective way to reduce the risk of mineral deficiency among populations, especially in developing countries where unrefined cereals and/or pulses are highly consumed (Kumar et al., 2010). Unfortunately, it is also associated with proliferation of microorganisms such as yeast and molds that may cause food safety concerns (Omemu, 2011). Salar et al. (2017) stated that fermentation of pearl millet with *Aspergillus sojae* for improvement of phenolic content, condensed tannin content, antioxidant potential and DNA damage protection activity may prove to be an important process for industrial usage.

2.6 POPPING

Puffing or popping is a traditional method used for producing ready-to-eat and stable shelf-life products that are crunchy and porous (Dutta et al., 2015). Puffing also involves soaking whole unhusked grains in water and mixing with sand heated at 250°C for 15–60 s (Sarkar et al., 2015). This process results in the development of a highly desirable aroma because of the Millard reaction between sugars and amino acids (Gull et al., 2016). Also, it can be pulverized and mixed with protein-rich sources to prepare ready nutritious supplementary food (Premavalli et al., 2003). Popping is a type of starch cookery that is the simplest, inexpensive, and fastest traditional method. It involves dry heat application on kernels of seeds until internal moisture expands, for preparation of healthy ready-to-eat snack products as well as making weaning foods (Chauhan & Sarita, 2018). Roopa and Premavalli (2008) reported that during puffing of ragi, RDS increased, with a decrease in SDS, which was more prominent in hilly varieties. However, the decrease in RDS was higher in the case of base varieties. Puffing increased gelatinization, thereby improving the starch digestibility.

2.7 SOAKING AND COOKING

Soaking of grains in plain water is a popular household food processing technique and a common practice to soften texture and hasten the cooking process (Silva et al., 1981). Soaking is a domestic technological treatment for the hydration of seeds in water for a few hours to allow the seeds to absorb water (Embaby, 2010). It is used for reducing antinutritional compounds such as phytic acid and phytase activity to improve bioavailability of minerals (Lestienne et al., 2005b). Similar to soaking, cooking in boiling water or steam also inactivates heat-labile antinutrients in ragi (Subhash et al., 2015). Jha et al. (2015) reported that flavanoid content decreased drastically by 50%–60% after the soaking process in pearl millets. This decrease could be due to the leaching of the pH-sensitive pigments in alkaline and acidic conditions and also due to steaming. Among the different soaking conditions, acidic soaking was effective in decreasing the phytate content in bran-rich fractions by about 81% and in the endosperm fraction by 75%. Lestienne et al. (2005) stated that depending on the botanical origin of the seeds, a significant reduction ($P < 0.05$) in phytate content (between 17% and 28%) was obtained by soaking whole seeds for 24 h at 30°C. Eyzaguirre et al. (2006) concluded that soaking of grains results in a 25% loss of iron but also facilitates endogenous phytate degradation, particularly when combined with milling and cooking. A study conducted by Lestienne et al. (2005c) showed that soaking resulted in reduction of phytate and zinc contents in pearl millet. Simple processing of foxtail millet like dehulling, soaking, and cooking resulted in a significant decrease in antinutrients, such as polyphenols and phytate, and improved the bioavailability of minerals such as iron and zinc and also protein digestibility in vitro (Pawar & Machewad, 2006). Pawar and Parlikar (1990) reported 18.52% and 74.07% reductions in phytate phosphorus on soaking pearl millet in water for 15 h and on dehulling (15.84% degree of dehulling), respectively. The foxtail millet sample soaked overnight resulted in the loss of iron and zinc caused by leaching out, and therefore the values for iron and zinc were decreased in the soaked grain samples than in the control one. During the water-soaking treatment, there was an increase in the ionizable iron and soluble zinc contents (Pawar & Machewad, 2006).

Hithamani and Srinivasan (2014) reported reduced total phenolic content, bioaccessible polyphenol, total flavonoids, bioaccessible flavonoids, and tannin along with increased % bioaccessibility of polyphenols and flavonoids after pressure cooking of finger millet (15 psi for 15–20 min). There are various factors that can influence the characteristics of the cooked grains, such as variety, type of pre-processing applied (e.g., decortication), grain size, water/grain ratio, cooking time, temperature profiles, and cooking method (Yu et al., 2009). Siroha and Sandhu (2017) observed that cooking for longer (time) decreases total phenolic content, total flavonoid content, antioxidant activity, and ABTS[+] scavenging activity, while the reverse is observed for metal chelating activity for pearl millet. Gulati et al. (2018) observed a significant decline in digestibility (more than 50%) of proso millet protein when it was heated above 55°C. The digestibility of proso millet proteins declines upon heating due to intramolecular hydrophobic protein aggregation (Gulati et al., 2017). Bora et al. (2019) reported that parboiling decreased the RDS in pearl and proso millet products from 18.2%–19.1% to 17.4%–18.3%, while the residual starch values increased from 44.4%–47.8% to 45%–49.1% and expected GI was decreased from 42.8–44.5 to 41.7–43.7. Chandrasekara et al. (2012) reported that cooked grains of millets (finger millet, proso, and little millets) had significantly ($P < 0.05$) lower DPPH radical scavenging activity than their dehulled uncooked counterparts. Dharmaraj et al. (2014) observed that a reduction in cooking time might be associated with various factors such as smaller size, removal of the seed coat, and a larger surface area, as well as the presence of pre-gelatinized starch in decorticated millet.

Microwave is an electro-heat technique that converts electrical energy into thermal energy on a frequency range of 300 MHz to 300 GHz, and food applications are in the range of 915 MHz and 2.45 GHz (Gavahian et al., 2019). Pradeep and Sreerama (2015) studied the effects of steam and microwave treatments, and observed that both steam and microwave treatments decreased the total phenolic content of barnyard millet, but increased the total phenolic content in foxtail and proso millets. However, steam treatment caused lower total flavonoid content levels in all the three millets.

The extrusion process is extensively applied in food industries to make breakfast cereals and snacks. Extrusion cooking is a high-temperature short-time cooking process that could be used for processing of starchy and proteinaceous materials (Jaybhaye et al., 2014). This processing also reduces the antinutritional factors, renders the product microbially safe, and enhances consumer acceptability (Nibedita & Sukumar, 2003). Flour from various plant sources can be prepared in different forms to produce common extruded products such as snacks, noodles, maca-roni, spaghetti, baby foods, and pasta, which are preferred by children and teenagers (Ramashia et al., 2019). Extrusion cooking works on the principle of combined efforts of shear force along with high pressure and temperature that is responsible for the modification of starch properties (Rathore et al., 2019).

2.8 ROASTING/TOASTING

Roasting/toasting is a rapid processing method that uses dry heat treatment for short time.

Roasting is a simple and commonly used household technology reported to improve the edibility and digestibility of grains, reduce their antinutrient levels, and prevent the loss of nutritious components (Huffman & Martin, 1994). Thermal treatments may reduce or increase the phenolic content and the antioxidant activities of cereals, depending on heat treatment, time period, and the type of cereal tested (Hegde & Chandra, 2005; Zieliński et al., 2006). Siroha and Sandhu (2017) observed that toasting significantly ($P < 0.05$) increases total phenolic content, total flavonoid content, antioxidant activity, and ABTS+ scavenging activity for pearl millet. Roasting reduced the protein, crude fiber, moisture content, lysine, methionine, potassium, phosphorus, calcium, magnesium, and phenolic contents of the pearl millet flour, their composition decreased as the roasting temperature increased. Roasting increased the ash, carbohydrate, energy value, and iron contents of the pearl millet flours but had no effect on the fat, threonine, and glycine contents (Obadina et al., 2016). Pradeep and Guha (2011) reported that roasting process increases the TPC, %DPPH inhibition, and reducing power assay of little millet. The increase in TPC during roasting may be due to the increase in the extractability of bound phenolics by the thermal degradation of cellular constituents. Hithamani and Srinivasan (2014) observed that after the roasting process, total polyphenols and bioaccessible flavonoids increased in amount, while bioaccessible polyphenols and total flavonoids content decreased. Roasting increases net protein utilization and enhances bioavailability of nutrients in roasted millets and legume mixes, which is superior to malted, dehulled, baked, and boiled mixes (Geervani et al., 1996; Gahlawat & Sehgal, 1994).

2.9 MALTING/GERMINATION

Germination is a biological process that is applied in order to obtain biotechnologically processed new food products (Jimenez et al., 2019). The application of controlled germination as a method of improving the nutritional value and flavor of grain products is of emerging interest. Reports have documented increases in reducing sugar and free amino acids (Ding et al., 2016), bioaccessible minerals (Platel et al., 2010), soluble dietary fiber (Koehler et al., 2007), phenolic compounds, and antioxidant capability (Hung et al., 2011) during grain germinating. The malting process, which involves soaking, germination, and drying, transforms grains into malt through high enzymatic activity (Amadou et al., 2011). Malting is an easily adaptable technology that causes increased activities of hydrolytic enzymes, leading to an increase in essential amino acids contents, total sugars, and B group vitamins, with a subsequent decrease in dry matter, starch, and antinutrients (Traoré et al., 2004; Coulibaly et al., 2012). During this process, the grains develop enzymes, including diastatic enzymes, which are required to modify the starch into sugars like monosaccharides (glucose or fructose) and disaccharides (sucrose or maltose). It also develops other enzymes, such as proteases, which break down the proteins in the grain (Banusha & Vasantharuba, 2013). Choi (1984) reported that starch in the endosperm was degraded slowly during the course of germination and, with the degradation of starch, the sugar levels were elevated during the period of germination. The rise in the level of reducing sugars may be due to mobilization and hydrolysis of seed polysaccharides, leading to more available reducing sugars. Sharma et al. (2018a) observed that germination of foxtail millet grains increased the protein, total dietary fiber, and total phenolic contents by 29.72%, 58.02%, and 77.42%, respectively, and ferrous ion-reducing antioxidant potential by 109.50% and reduced the fat content by 27.98%. This could be due to intrinsic biosynthesis of free and soluble forms of nutrients in germinated foxtail millet gain samples through enzymatic reaction during the germination process (Xia et al., 2017). Sharma et al. (2018b) reported that germination process increases the total antioxidant activity (29.0–45.23 mgAAE/g), reducing power activity (0.53–0.76 µg/ml), DPPH activity (48.32%–59.62%), and hydrogen peroxide scavenging activities (35.44–63.07 mM–Trolox/g) of foxtail millet. Li et al. (2017) reported that amylose content, relative crystallinity, lower retrogradation enthalpy, and peak viscosity of starches decreased after the germination process of millets. Sharma et al. (2016) reported that with an increase in soaking and germination time as well as germination temperature, the antioxidant activity, total phenolic component, and total flavanoid component of barnyard millet increased significantly. Malleshi and Desikachar (1986) stated increase in lysine and methionine contents from 3.5 to 4.0 mg/100 and 1.3 to 1.5 mg/100 g, respectively, in germinated finger millet. Malting also increases threonine, tryptophan, and protein contents of finger millets. Proteases are activated during malting, causing the protein to degrade and bioavailability to improve (Pyler & Thomas, 2000). The malting process decreases phytic acid and increases the minerals' extractability (Lestienne et al., 2005). This process has also been reported to enhance flavor and bioactivity of grains (Heinio et al., 2001). Hithamani and Srinivasan (2014) reported that sprouting did not alter the polyphenol content, while bioaccessibility of identified phenolic compounds increased by 40% on sprouting. Total flavonoid content of the grains reduced drastically on sprouting. Total soluble sugar levels, including reducing and non-reducing sugar, of the germinated pearl millet were reported to be higher than the control sample (Khetarpaul & Chauhan, 1990). The sprouting process decreased the protein, fat, starch, ash contents, dietary fibers, amylase, and protease activity (Sharma & Gujral, 2019). Sprouting promoted a significant ($P < 0.05$) decrease in antinutritional components like phytic acid and condensed tannin content (Figure 2.2).

2.10 CONCLUSION

Processing technologies can be used to improve nutrient content and also play a role in enhancing the dietary quality of millets. Generally, processing alters the grain quality, improves the nutritional availability and storage stability of the flour as well as the products themselves, and helps retain all parts of the whole grains that are beneficial for health and pleasant to consume. Different processing methods such as fermentation, malting, decortications, roasting, and toasting are used for millets. There is, however, need for extensive research to formulate the product from processed millet grains that provides good nutritional properties with improved sensorial properties, including taste, aroma, color, and texture.

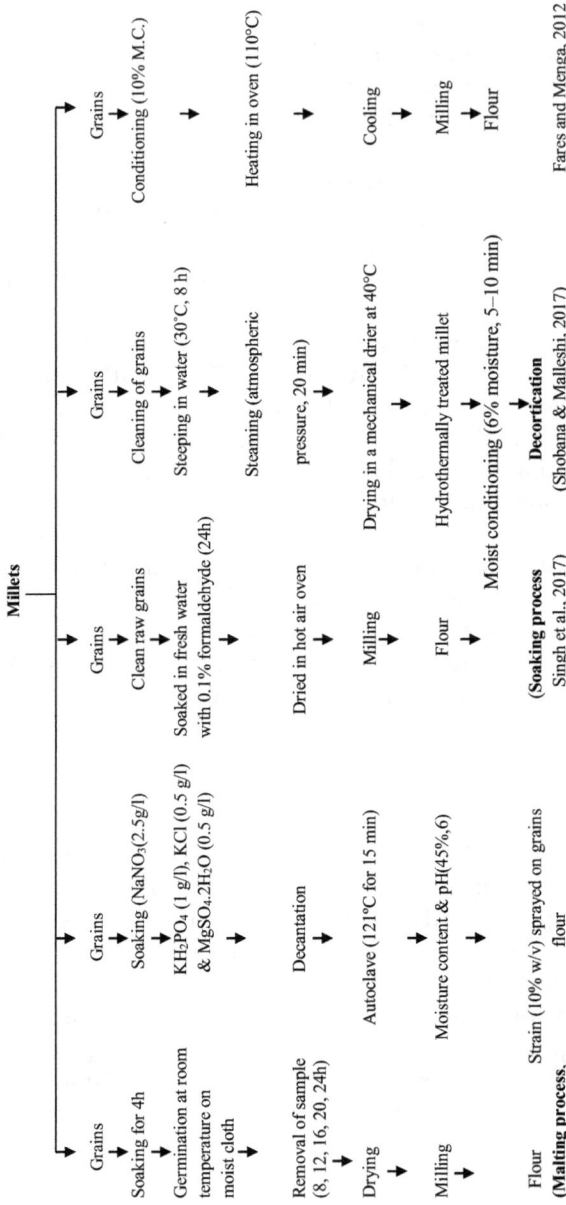

Figure 2.2 Different methods of processing.

REFERENCES

Abdel-Haleem, W. H., El Tinay, A. H., Mustafa, A. I. and Babiker, E. E. 2008. Effect of fermentation, malt-pretreatment and cooking on antinutritional factors and protein digestibility of sorghum cultivars. *Pakistan Journal of Nutrition* 7: 335–341.

Adebiyi, J. A., Obadina, A. O., Adebo, O. A. and Kayitesi, E. 2018. Fermented and malted millet products in Africa: expedition from traditional/ethnic foods to industrial value-added products. *Critical Reviews in Food Science and Nutrition* 58: 463–474.

Adebiyi, J. A., Obadina, A. O., Mulaba-Bafubiandi, A. F., Adebo, O. A. and Kayitesi, E. 2016. Effect of fermentation and malting on the microstructure and selected physicochemical properties of pearl millet (*Pennisetum glaucum*) flour and biscuit. *Journal of Cereal Science* 70: 132–139.

Agte, V. V., Gokhale, M. K. and Chiplonkar, S. A. 1997. Effect of natural fermentation on in vitro zinc bioavailability in cereal–legume mixtures. *International Journal of Food Science & Technology* 32: 29–32.

Ali, M. A., El Tinay, A. H. and Abdalla, A. H. 2003. Effect of fermentation on the in vitro protein digestibility of pearl millet. *Food Chemistry* 80: 51–54.

Amadou, I., Gbadamosi, O. S. and Le, G. W. 2011. Millet-based traditional processed foods and beverages–a review. *Cereal Foods World* 56: 115–121.

Amadou, I., Gounga, M. E., Shi, Y. H. and Le, G. W. 2014. Fermentation and heat-moisture treatment induced changes on the physicochemical properties of foxtail millet (*Setaria italica*) flour. *Food and Bioproducts Processing* 92: 38–45.

Amadou, I., Le, G. W., Amza, T., Sun, J. and Shi, Y. H. 2013b. Purification and characterization of foxtail millet-derived peptides with antioxidant and antimicrobial activities. *Food Research International* 51: 422–428.

Amadou, I., Le, G. W. and Shi, Y. H. 2013a. Evaluation of antimicrobial, antioxidant activities, and nutritional values of fermented foxtail millet extracts by *Lactobacillus paracasei* Fn032. *International Journal of Food Properties* 16: 1179–1190.

Antony, U. and Chandra, T. S. 1998. Antinutrient reduction and enhancement in protein, starch and mineral availability in fermented flour of finger millet (*Eleusine coracana*). *Journal of Agricultural and Food Chemistry* 46: 2578–2582.

Ayo, J. A. and Olawale, O. 2003. Effect of defatted groundnut concentrate on the physico-chemical and sensory quality of fura. *Nutrition and Food Science* 33: 173–176.

Banusha, S. and Vasantharuba, S. 2013. Effect of malting on nutritional contents of finger millet and mung bean. *American-Eurasian Journal of Agriculture & Environmental Science* 13: 1642–1646.

Bender, D. A. 2006. *Bender's dictionary of nutrition and food technology*. 8th ed. Abingdon, UK: Woodhead Publishing & CRC Press.

Bora, P., Ragaee, S. and Marcone, M. 2019. Effect of parboiling on decortication yield of millet grains and phenolic acids and in vitro digestibility of selected millet products. *Food Chemistry* 274: 718–725.

Chandrasekara, A., Naczk, M. and Shahidi, F. 2012. Effect of processing on the antioxidant activity of millet grains. *Food Chemistry* 133: 1–9.

Chauhan, E. S. and Sarita 2018. Effects of processing (germination and popping) on the nutritional and anti-nutritional properties of finger millet (*Eleusine Coracana*). *Current Research in Nutrition and Food Science Journal* 6: 566–572.

Chavan, J. K., Kadam, S. S. and Beuchat, L. R. 1989. Nutritional improvement of cereals by sprouting. *Critical Reviews in Food Science & Nutrition* 28: 401–437.

Choi, K. C. 1984. Amylase activity during germination of maize seeds. *Journal of the Korean Agricultural Chemical Society* 27: 107–11.

Coulibaly, A., Kouakou, B. and Chen, J. 2012. Extruded adult breakfast based on millet and soybean: nutritional and functional qualities, source of low glycemic food. *Nutrition & Food Sciences* 2: 1–9.

Davidek, J., Velisek, J. and Pokorny, J. 1990. *Chemical changes during food processing.* New York, Tokyo, Elsevier Amsterdam-Oxford. p. 448.

Dharmaraj, U., Ravi, R. and Malleshi, N. G. 2014. Cooking characteristics and sensory qualities of decorticated finger millet (*eleusine coracana*). *Journal of Culinary Science & Technology* 12: 215–228.

Dias-Martins, A. M., Cappato, L. P., da Costa Mattos, M., Rodrigues, F. N., Pacheco, S. and Carvalho, C. W. 2019. Impacts of ohmic heating on decorticated and whole pearl millet grains compared to open-pan cooking. *Journal of Cereal Science* 85: 120–129.

Dias-Martins, A. M., Pessanha, K. L. F., Pacheco, S., Rodrigues, J. A. S. and Carvalho, C. W. P. 2018. Potential use of pearl millet (*Pennisetum glaucum* (L.) R. Br.) in Brazil: food security, processing, health benefits and nutritional products. *Food Research International* 109: 175–186.

Ding, J., Yang, T., Feng, H., Dong, M., Slavin, M., Xiong, S. and Zhao, S. 2016. Enhancing contents of γ-amino butyric acid (GABA) and other micronutrients in dehulled rice during germination under normoxic and hypoxic conditions. *Journal of Agricultural and Food Chemistry* 64: 1094–1102.

Divate, A., Sawant, A. A. and Thakor, N. J. 2015. Effect of extruder temperature on functional characteristics of finger millet (*Eleusine coracana* (L) Gaertn) based extrudates. *Global Science Research Journals* 3: 239–246.

Dutta, A., Mukherjee, R., Gupta, A., Ledda, A. and Chakraborty, R. 2015. Ultrastructural and physicochemical characteristics of rice under various conditions of puffing. *Journal of Food Science and Technology* 52: 7037–7047.

Ejigui, J., Savoie, L., Marin, J. and Desrosiers, T. 2005. Beneficial changes and drawbacks of a traditional fermentation process on chemical composition and anti-nutritional factors of yellow maize (*Zea mays*). *Journal of Biological Sciences* 5: 590–596.

Eltayeb, M. M., Hassan, A. B., Sulieman, M. A. and Babiker, E. E. 2007. Effect of processing followed by fermentation on antinutritional factors content of pearl millet (*Pennisetum glaucum* L.) cultivars. *Pakistan Journal of Nutrition* 6: 463–467.

44

El-Tinay, A. H., Abdel Gadir, A. M. and El-Hidi, M. 1979. Sorghum fermented kisra bread. I—nutritive value of kisra. *Journal of the Science of Food and Agriculture* 30: 859–863.

Elyas, S. H., El Tinay, A. H., Yousif, N. E. and Elsheikh, E. A. 2002. Effect of natural fermentation on nutritive value and in vitro protein digestibility of pearl millet. *Food Chemistry* 78: 75–79.

Embaby, H. E. S. 2010. Effect of heat treatments on certain antinutrients and in vitro protein digestibility of peanut and sesame seeds. *Food Science and Technology Research* 17: 31–38.

Etcheverry, P., Grusak, M. A. and Fleige, L. E. 2012. Application of in vitro bioaccessibility and bioavailability methods for calcium, carotenoids, folate, iron, magnesium, polyphenols, zinc, and vitamins B6, B12, D, and E. *Frontiers in Physiology* 3: 1–22.

Eyzaguirre, R. Z., Nienaltowska, K., De Jong, L. E., Hasenack, B. B. and Nout, M. R. 2006. Effect of food processing of pearl millet (*Pennisetum glaucum*) IKMP-5 on the level of phenolics, phytate, iron and zinc. *Journal of the Science of Food and Agriculture* 86: 1391–1398.

Farooq, U., Mohsin, M., Liu, X. and Zhang, H. 2013. Enhancement of short chain fatty acid production from millet fibres by pure cultures of probiotic fermentation. *Tropical Journal of Pharmaceutical Research* 12: 189–194.

Gahlawat, P. and Sehgal, S. 1994. Protein and starch digestibility and iron availability in developed weaning foods as affected by roasting. *Journal of Human Nutrition & Dietetics* 7: 121–126.

Gavahian, M., Chu, Y. H. and Farahnaky, A. 2019. Effects of ohmic and microwave cooking on textural softening and physical properties of rice. *Journal of Food Engineering* 243: 114–124.

Geervani, P., Vimala, V., Pradeep, K. U. and Devi, M. R. 1996. Effect of processing on protein digestibility, biological value and net protein utilization of millet and legume based infant mixes and biscuits. *Plant Foods for Human Nutrition* 49: 221–227.

Gulati, P., Li, A., Holding, D., Santra, D., Zhang, Y. and Rose, D. J. 2017. Heating reduces proso millet protein digestibility via formation of hydrophobic aggregates. *Journal of Agricultural & Food Chemistry* 65: 1952–1959.

Gulati, P., Zhou, Y., Elowsky, C. and Rose, D. J. 2018. Microstructural changes to proso millet protein bodies upon cooking and digestion. *Journal of Cereal Science* 80: 80–86.

Gull, A., Ahmad, N. G., Prasad, K. and Kumar, P. 2016. Technological, processing and nutritional approach of finger millet (*Eleusine coracana*)—a mini review. *Journal of Food Processing & Technology* 7: 1–4.

Hadimani, N. A. and Malleshi, N. G. 1993. Studies on milling, physico-chemical properties, nutrient composition and dietary fibre content of millets. *Journal of Food Science & Technology* 30: 17–20.

Hag, M. E. E., Tinay, A. H. E. and Yousif, N. E. 2002. Effect of fermentation and dehulling on starch, total polyphenols, phytic acid content and *in vitro* protein digestibility of pearl millet. *Food Chemistry* 77: 193–196.

Hama, F., Icard-Vernière, C., Guyot, J. P., Picq, C., Diawara, B. and Mouquet-Rivier, C. 2011. Changes in micro-and macronutrient composition of pearl millet and white sorghum during in field versus laboratory decortication. *Journal of Cereal Science* 54: 425–433.

Hegde, P. S. and Chandra, T. S. 2005. ESR spectroscopic study reveals higher free radical quenching potential in kodo millet (*Paspalum scrobiculatum*) compared to other millets. *Food Chemistry* 92: 177–182.

Heinio, R. L., Katina, K., Wilhelmson, A., Myllymaki, O., Rajamaki, T. and Latva-Kala, K. 2003. Relationship between sensory perception and flavour-active volatile compounds of germinated, sourdough fermented and native rye following the extrusion process. *LWT-Food Science & Technology* 36: 533–545.

Heinio, R. L., Oksman-Caldentey, K. M., Latva-Kala, K., Lehtinen, P. and Poutanen, K. 2001. Effects of drying treatment conditions on sensory profile of germinated oat. *Cereal Chemistry* 78: 707–714.

Hithamani, G. and Srinivasan, K. 2014. Effect of domestic processing on the polyphenol content and bioaccessibility in finger millet (*Eleusine coracana*) and pearl millet (*Pennisetum glaucum*). *Food Chemistry* 164: 55–62.

Hole, A. S., Rud, I., Grimmer, S., Sigle, S., Narvhus, J. and Sahlstrøm, S. 2012. Improved bioavailability of dietary phenolic acids in whole grain barley and oat groat following fermentation with probiotic *Lactobacillus acidophilus*, *Lactobacillus johnsonii*, and *Lactobacillus reuteri*. *Journal of Agricultural and Food Chemistry* 60: 6369–6375.

Hotz, C. and Gibson, R. S. 2007. Traditional food-processing and preparation practices to enhance the bioavailability of micronutrients in plant-based diets. *The Journal of Nutrition* 137: 1097–1100.

Huffman, S. L. and Martin, L. H. 1994. First feedings: optimal feeding of infant and toddlers. *Nutrition Review* 14: 127–159.

Hung, P. Van, Hatcher, D. W. and Barker, W. 2011. Phenolic acid composition of sprouted wheats by ultra-performance liquid chromatography (UPLC) and their antioxidant activities. *Food Chemistry* 126: 1896–1901.

Inyang, C. U. and Zakari, U. M. 2008. Effect of germination and fermentation of pearl millet on proximate, chemical and sensory properties of instant *"fura"*–a Nigerian cereal food. *Pakistan Journal of Nutrition* 7: 9–12.

Jay, J. M., Loessner, M. J. and Golden, D. A. 2005. *Modern food microbiology*. 7th ed. India: Springer.

Jaybhaye, R. V., Pardeshi, I. L., Vengaiah, P. C. and Srivastav, P. P. 2014. Processing and technology for millet based food products: a review. *Journal of Ready to Eat Food* 1: 32–48.

Jha, N., Krishnan, R. and Meera, M. S. 2015. Effect of different soaking conditions on inhibitory factors and bioaccessibility of iron and zinc in pearl millet. *Journal of Cereal Science* 66: 46–52.

Jimenez, M. D., Lobo, M. and Sammán, N. 2019. 12th IFDC 2017 special issue–influence of germination of quinoa (*Chenopodiu quinoa*) and amaranth (*Amaranthus*) grains on nutritional and techno-functional properties of their flours. *Journal of Food Composition and Analysis* 84: 103290.

Katina, K., Liukkonen, K. H., Kaukovirta-Norja, A., Adlercreutz, H., Heinonen, S. M. and Lampi, A. M. 2007. Fermentation-induced changes in the nutritional value of native or germinated rye. *Journal of Cereal Science* 46: 348–355.

Khetarpaul, N. and Chauhan, B. M. 1990. Effect of germination and fermentation on available carhohydrate content of pearl millet. *Food Chemistry* 38: 21–26.

Koehler, P., Hartmann, G., Wieser, H. and Rychlik, M. 2007. Changes of folates, dietary fiber, and proteins in wheat as affected by germination. *Journal of Agricultural & Food Chemistry* 55: 4678–4683.

Kulkarni, D. B., Sakhale, B. K. and Giri, N. A. 2018. A potential review on millet grain processing. *International Journal of Nutritional Sciences*: 1–8.

Kumar, V., Sinha, A. K., Makkar, H. P. S. and Becker, K. 2010. Dietary roles of phytate and phytase in human nutrition: a review. *Food Chemistry* 120: 945–959.

Kurien, P. P., Joseph, K., Swaminathan, M., Subrahmanyan, V. and Daniel, V. A. 1959. The distribution of nitrogen, calcium and phosphorus between the husk and endosperm of Ragi (*Eleusine coracana*). *Food Science* 8: 353–355.

Lestienne, I., Besancon, P., Caporiccio, B., Lullien-Pellerin, V. and Treche, S. 2005a. Iron and zinc in vitro availability in pearl millet flours (*Pennisetum glaucum*) with varying phytate, tannin, and fiber contents. *Journal Agricultural Food Chemistry* 53: 3240–3247.

Lestienne, I., Icard-Vernière, C., Mouquet, C., Picq, C. and Trèche, S. 2005b. Effects of soaking whole cereal and legume seeds on iron, zinc and phytate contents. *Food Chemistry* 89: 421–425.

Lestienne, I., Mouquet-Rivier, C., Icard-Vernière, C., Rochette, I. and Trèche, S. 2005c. The effects of soaking of whole, dehulled and ground millet and soybean seeds on phytate degradation and Phy/Fe and Phy/Zn molar ratios. *International Journal of Food Science & Technology* 40: 391–399.

Li, C., Oh, S. G., Lee, D. H., Baik, H. W. and Chung, H. J. 2017. Effect of germination on the structures and physicochemical properties of starches from brown rice, oat, sorghum, and millet. *International Journal of Biological Macromolecules* 105: 931–939.

Liu, J., Tang, X., Zhang, Y. and Zhao, W. 2012. Determination of the volatile composition in brown millet, milled millet and millet bran by gas chromatography/mass spectrometry. *Molecules* 17: 2271–2282.

Mal, B., Padulosi, S. and Ravi, S. B. 2010. Minor millets in South Asia: learnings from IFAD-NUS project in India and Nepal. Maccarese, Rome, Italy: Bioversity Intl., and Chennai, India: M.S. Swaminathan Research Foundation: 1–185.

Malleshi, N. G. and Desikachar, H. S. R. 1986. Influence of malting conditions on quality of finger millet. *Journal of Industrial Brewing* 92: 81.

Mohamed, E. A., Ali, N. A., Ahmed, S. H., Ahmed, I. A. M. and Babiker, E. E. 2010. Effect of radiation process on antinutrients and HCl extractability of calcium, phosphorus and iron during processing and storage. *Radiation Physics & Chemistry* 79: 791–796.

Nantanga, K. K., Seetharaman, K., de Kock, H. L. and Taylor, J. 2008. Thermal treatments to partially pre-cook and improve the shelf-life of whole pearl millet flour. *Journal of the Science of Food & Agriculture* 88: 1892–1899.

Nazni, P. and Devi, S. R. 2016. Effect of processing on the characteristics changes in barnyard and foxtail millet. *Journal of Food Processing and Technology* 7: 1–9.

Nibedita, M. and Sukumar, B. 2003. Extrusion cooking technology employed to reduce the anti-nutritional factor tannin in sesame (*Sesamum indicum*) meal. *Journal of Food Engineering* 56: 201–202.

Nithya, K. S., Ramachandramurty, B. and Krishnamoorthy, V. V. 2007. Effect of processing methods on nutritional and anti-nutritional qualities of hybrid (COHCU-8) and traditional (CO7) pearl millet varieties of India. *Journal of Biological Sciences* 7: 643–647.

Obadina, A., Ishola, I. O., Adekoya, I. O., Soares, A. G., de Carvalho, C. W. P. and Barboza, H. T. 2016. Nutritional and physico-chemical properties of flour from native and roasted whole grain pearl millet (*Pennisetum glaucum* [L.] R. Br.). *Journal of Cereal Science* 70: 247–252.

Omemu, A. M. 2011. Fermentation dynamics during production of ogi, a Nigerian fermented cereal porridge. *Report and Opinion* 3: 8–17.

Osman, M. A. 2004. Changes, in sorghum enzyme inhibitors, phytic acid, tannins, and in vitro protein digestibility occurring during Khamir (local bread) fermentation. *Food Chemistry* 88: 129–134.

Osman, M. A. 2011. Effect of traditional fermentation process on the nutrient and antinutrient contents of pearl millet during preparation of Lohoh. *Journal of the Saudi Society of Agricultural Sciences* 10: 1–6.

Pampangouda, P., Munishamanna, K. B. and Gurumurthy, H. 2015. Effect of *Saccharomyces boulardii* and *Lactobacillus acidophilus* fermentation on little millet (*Panicum sumatrense*). *Journal of Applied & Natural Science* 7: 260–264.

Pawar, V. D. and Machewad, G. M. 2006. Processing of foxtail millet for improved nutrient availability. *Journal of Food Processing & Preservation* 30: 269–279.

Pawar, V. D. and Parlikar, G. S. 1990. Reducing the polyphenols and phytate and improving the protein quality of pearl millet by dehulling and soaking. *Journal of Food Science & Technology* 27: 140–143.

Pellegrini, N., Chiavaro, E., Gardana, C., Mazzeo, T., Contino, D., et al. 2010. Effect of different cooking methods on color, phytochemical concentration, and antioxidant capacity of raw and frozen brassica vegetables. *Journal of Agricultural & Food Chemistry* 58: 4310–4321.

Platel, K., Eipeson, S. W. and Srinivasan, K. 2010. Bioaccessible mineral content of malted finger millet (*Eleusine coracana*), wheat (*Triticum aestivum*), and barley (*Hordeum vulgare*). *Journal of Agricultural & Food Chemistry* 58: 8100–8103.

Pradeep, S. R. and Guha, M. 2011. Effect of processing methods on the nutraceutical and antioxidant properties of little millet (*Panicum sumatrense*) extracts. *Food Chemistry* 126: 1643–1647.

Pradeep, P. M. and Sreerama, Y. N. 2015. Impact of processing on the phenolic profiles of small millets: evaluation of their antioxidant and enzyme inhibitory properties associated with hyperglycemia. *Food chemistry* 169: 455–463.

Premavalli, K. S., Majumdar, T. K., Madhura, C. V. and Bawa, A. S. 2003. Development of traditional products. V. Ragi based convenience mixes. *Journal of Food Science & Technology* 40: 361–365.

Pyler, R. E. and Thomas, D. A. 2000. Malted cereals: their production and use. In: Kulp, K. and Ponte, J. G., editors. *Handbook of cereal science and technology*. New York: Marcel Dekker.

Rai, K. N., Gowda, C. L. L., Reddy, B. V. S. and Sehgal, S. 2008. Adaptation and potential uses of sorghum and pearl millet in alternative and health foods. *Comprehensive Reviews in Food Science & Food Safety* 7: 340–352.

Ramashia, S. E., Anyasi, T. A., Gwata, E. T., Meddows-Taylor, S. and Jideani, A. I. O. 2019. Processing, nutritional composition and health benefits of finger millet in sub-saharan Africa. *Food Science & Technology* 39: 253–266.

Rathore, T., Singh, R., Kamble, D. B., Upadhyay, A. and Thangalakshmi, S. 2019. Review on finger millet: processing and value addition. *The Pharma Innovation Journal* 8: 283–291.

Rathore, S., Singh, K. and Kumar, V. 2016. Millet grain processing, utilization and its role in health promotion – a review. *International Journal of Nutrition & Food Sciences* 5: 318–329.

Roopa, S. and Premavalli, K. S. 2008. Effect of processing on starch fractions in different varieties of finger millet. *Food Chemistry* 106: 875–882.

Salar, R. K. and Purewal, S. S. 2016. Improvement of DNA damage protection and antioxidant activity of biotrans formed pearl millet (*Pennisetum glaucum*) cultivar PUSA-415 using *Aspergillus oryzae* MTCC 3107. *Biocatalysis & Agricultural Biotechnology* 8: 221–227.

Salar, R. K., Purewal, S. S. and Sandhu, K. S. 2017. Fermented pearl millet (*Pennisetum glaucum*) with *in vitro* DNA damage protection activity, bioactive compounds and antioxidant potential. *Food Research International* 100: 204–210.

Saleh, A. S., Zhang, Q., Chen, J. and Shen, Q. 2013. Millet grains: nutritional quality, processing, and potential health benefits. *Comprehensive Reviews in Food Science & Food Safety* 12: 281–295.

Sandhu, K. S., Punia, S. and Kaur, M. 2017. Fermentation of cereals: a tool to enhance bioactive compounds. In: Gahlawat, S. K., Salar, R. K., Siwach, P., Duhan, J. S., Kumar, S. and Kaur, P., editors. *Plant biotechnology: recent advancements and developments*. Singapore: Springer, pp. 157–170.

Sarkar, P., Lohith Kumar, D. H., Dhumal, C., Panigrahi, S. S. and Choudhary, R. 2015. Traditional and ayurvedic foods of Indian origin. *Journal of Ethnic Foods* 2: 97–109.

Serna-Saldivar, S. O., Clegg, C. and Rooney, L. W. 1994. Effects of parboiling and decortication on the nutritional value of sorghum (*Sorghum bicolor* L. *Moench*) and pearl millet (*Pennisetum glaucum* L.). *Journal of Cereal Science* 19: 83–89.

Sharma, N., Goyal, S. K., Alam, T., Fatma, S., Chaoruangrit, A. and Niranjan, K. 2018a. Effect of high pressure soaking on water absorption, gelatinization, and biochemical properties of germinated and non-germinated foxtail millet grains. *Journal of Cereal Science* 83: 162–170.

49

Sharma, B. and Gujral, H. S. 2019. Modifying the dough mixing behavior, protein and starch digestibility and antinutritional profile of minor millets by sprouting. *International Journal of Biological Macromolecules* 153: 962–970.

Sharma, P., Gujral, H. S. and Singh, B. 2012. Antioxidant activity of barley as affected by extrusion cooking. *Food Chemistry* 131: 1406–1413.

Sharma, S., Saxena, D. C. and Riar, C. S. 2016. Analysing the effect of germination on phenolics, dietary fibres, minerals and γ-amino butyric acid contents of barnyard millet (*Echinochloa frumentaceae*). *Food Bioscience* 13: 60–68.

Sharma, S., Saxena, D. C. and Riar, C. S. 2018b. Changes in the GABA and polyphenols contents of foxtail millet on germination and their relationship with *in vitro* antioxidant activity. *Food Chemistry* 245: 863–870.

Shobana, S. and Malleshi, N. G. 2007. Preparation and functional properties of decorticated finger millet (*Eleusine coracana*). *Journal of Food Engineering* 79: 529–538.

Silva, C. A. B., Rayes, R. P. and Deng, J. C. 1981. Influence of soaking and cooking upon softening and eating quality of black bean (*Phaseolus vulgaris*). *Journal of Food Science* 46: 1716–1720.

Singh, K. P. 2010. Development of a dehuller for barnyard millet (*Echinochloa frumentacea*) and formulation of millet-wheat composite flour. Unpublished Ph. D. Thesis, Agriculture and Food Engineering Department, IIT, Kharagpur (W.B.)-721302, India.

Singh, P. and Raghuvanshi, R. S. 2012. Finger millet for food and nutritional security. *African Journal of Food Science* 6: 77–84.

Singh, A. K., Rehal, J., Kaur, A. and Jyot, G. 2015. Enhancement of attributes of cereals by germination and fermentation: a review. *Critical Reviews in Food Science & Nutrition* 55: 1575–1589.

Siroha, A. K. and Sandhu, K. S. 2017. Effect of heat processing on the antioxidant properties of pearl millet (*Pennisetum glaucum* L.) cultivars. *Journal of Food Measurement & Characterization* 11: 872–878.

Subhash, B., Kakade and Hathan, B. S. 2015. Finger millet processing: review. *International Journal of Agriculture Innovations & Research* 3: 1003–1008.

Traoré, T., Mouquet, C., Icard-Vernière, C., Traoré, A. and Trèche, S. 2004. Changes in nutrient composition, phytate and cyanide contents and α-amylase activity during cereal malting in small production units in Ouagadougou (*Burkina Faso*). *Food Chemistry* 88: 105–114.

Ushakumari, S. R., Shrikantan, L. and Malleshi, N. G. 2004. The functional properties of popped, flaked, extruded and roller dried foxtail millet (*Setaria italica*). *International Journal of Food Science & Technology* 39: 907–915.

Verma, V. and Patel, S. 2013. Value added products from nutria-cereals, finger millet (*Eleusine coracana*). *Emirates Journal of Food & Agriculture* 25: 169–176.

Xia, Q., Wang, L., Xu, C., Mei, J. and Li, Y. 2017. Effects of germination and high hydrostatic pressure processing on mineral elements, amino acids and antioxidants in vitro bioaccessibility, as well as starch digestibility in brown rice (*Oryza sativa* L.). *Food Chemistry* 214: 533–542.

Yadav, D. N., Anand, T., Kaur, J. and Singh, A. K. 2012. Improved storage stability of pearl millet flour through microwave treatment. *Agricultural Research* 1: 399–404.

Young, R., Haidara, M., Rooney, L. W. and Waniska, R. D. 1990. Parboiled sorghum: development of a novel decorticated product. *Journal of Cereal Science* 11: 277–289.

Yu, S., Ma, Y. and Sun, D. W. 2009. Impact of amylose content on starch retrogradationand texture of cooked milled rice during storage. *Journal Cereal Science* 50: 139–144

Zieliński, H., Michalska, A., Piskuła, M. K. and Kozłowska, H. 2006. Antioxidants in thermally treated buckwheat groats. *Molecular Nutrition & Food Research* 50: 824–832.

3

Physical and Functional Properties of Millets

3.1 INTRODUCTION

Physical properties are indicators of the quality of cereal grains. Equipment for the handling, harvesting, processing, sorting (Davies, 2009), cleaning, grading, and separation (Baryeh, 2001) are often designed on the basis of these properties. In the designing of silos, bins, hoppers, and storage structures, the bulk density, true density, porosity, and angle of repose are important parameters. Projected area and volume values of the grains are important parameters in cooking, aeration, and artificial drying processes (Kheiralipour et al., 2008). The physical properties are also important for developing sensors for automation of machineries and processes. Esref and Halil (2007) acknowledged the importance of physical properties for storage structures and processing operations. The irregular shapes and composition of most of the biological materials further signify the utility of these characteristics.

Functional properties are the essential physicochemical properties of foods that reflect the complex interactions between the structures, molecular conformation, compositions, and physicochemical properties of food components with the nature of the environment and conditions in which these are measured and associated (Chandra & Singh, 2013; Siddiq et al., 2009). The use of grain flours in food formulations is dependent on flour functionality. The water absorption capacity plays an important role in the development of food products because it influences, to a large extent, their interaction with water. The protein/nitrogen solubility value provides useful information on effective utilization (Devisetti et al., 2014).

Physicochemical and nutritional properties of major millets make them suitable for large-scale utilization and consumption in the world, by manufacturing a wide variety of value-added food products, e.g. baby foods, dietary foods, and ready-to-eat (RTE) snacks (Subramanian & Viswanathan, 2003). When millets are utilized in the production of extrusion-cooked and baked products, they are in the form of flour. Hence, it also becomes essential to study the physical properties of the flour (Teunou & Fitzpatrick, 1999). Functional properties (water absorption capacity, oil absorption capacity, foaming capacity and stability, emulsion capacity, and stability) play important roles in the manufacturing of products and affect the sensory characteristics of foods.

3.2 PHYSICAL PROPERTIES

Millet grains available on the market are of varying sizes, shapes, and colors (Vadivoo et al., 1998). Varnamkhasti et al. (2008) demonstrated that physical properties of grains include 1000 kernel weight, moisture content, porosity, bulk density, sample volume, aspect ratio, true density, sample surface area, and perpendicular dimensions (length, width, and thickness) (Figure 3.1). Kumari and Raghuvanshi (2015) demonstrated that functional properties include viscosity, foaming capacity (FC), water absorption capacity (WAC), dispersibility, oil absorption capacity (OAC), and swelling capacity (SC) (Figure 3.2). The major equipment that are useful during the determination of physical and functional properties of millet grains are reported in Figure 3.3. Faleye et al. (2013) observed that a detailed study of grains in terms of their physical and functional parameters helps to prepare specific food products. Detailed knowledge of physical and functional properties of grains also helps in product formulation, harvesting, and storage under different types of conditions. Further, knowledge of physical properties helps in machinery designing for harvest. Physical properties of agricultural materials are important for designing appropriate equipment and systems for planting, harvesting, and postharvest operations such as cleaning, conveyance, and storage (Asoegwu et al., 2006). The physical and mechanical properties of the flours of different millets are summarized in Table 3.1.

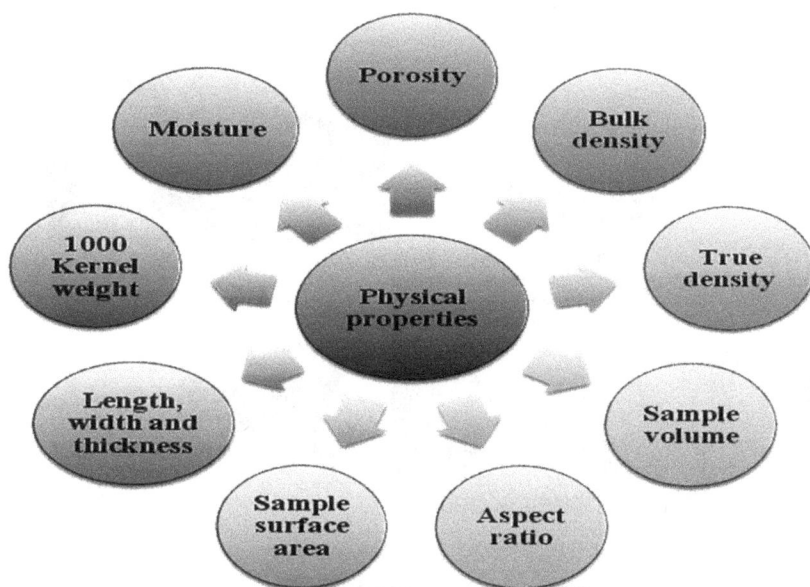

Figure 3.1 Specific physical properties.

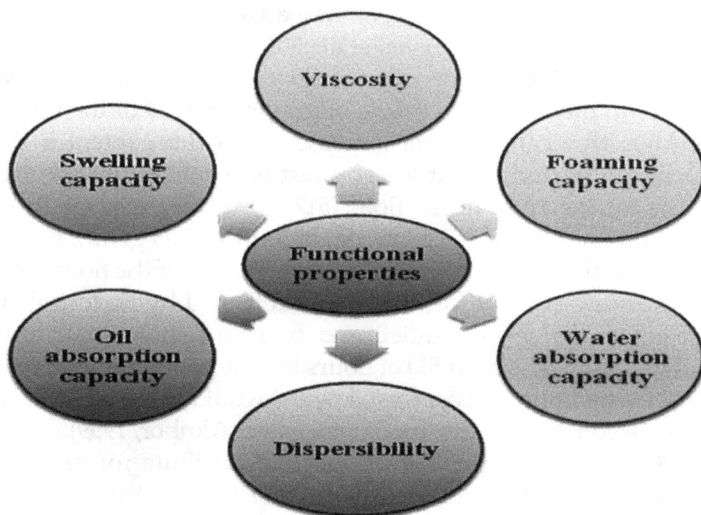

Figure 3.2 Functional properties.

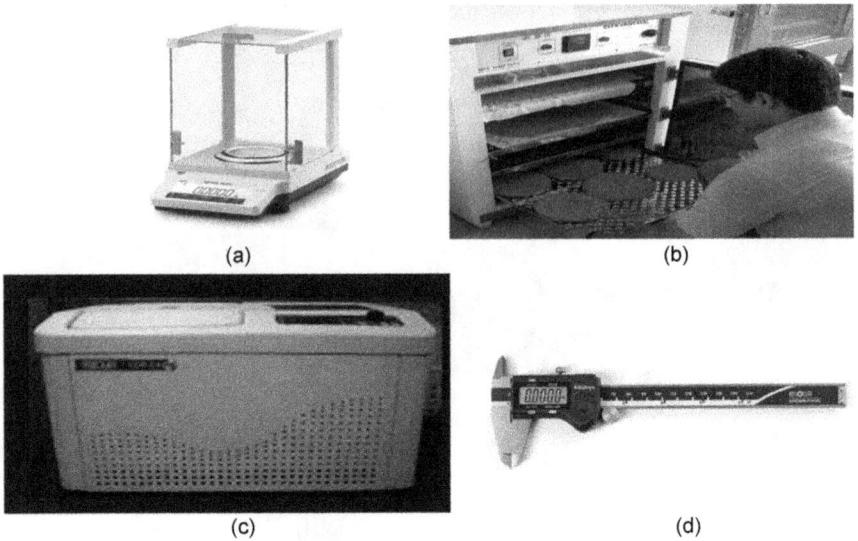

Figure 3.3 Equipment required for determination of physical and functional properties: (a) weighing balance, (b) hot air oven, (c) centrifuge, and (d) Vernier caliper.

Bulk density (BD) is a reflection of the load a sample can carry if allowed to rest directly on another (Kaushal et al., 2012). The BDs of different millet grains are shown Table 3.1. The value of BDs of millet grains was decreased with increase in the moisture content of grains. The BD of the millet flours increased logarithmically with the increase in moisture content, in the moisture content range studied. At lower moisture content, BD was observed to be higher for the kodo millet flour (517 kg/m³) followed by little millet (515 kg/m³), proso millet (505 kg/m³), foxtail millet (503 g/m³), finger millet (490 kg/m³), and barnyard millet (485 kg/m³). The BD of the flour was higher at higher moisture content for little millet followed by kodo millet, proso millet, foxtail millet, finger millet, and barnyard millet (Subramanian& Viswanathan, 2007). The high BD of flours thus supports their suitability for use in food preparations. In contrast, low BD would be an advantage in the formulation of complementary foods (Akpata & Akubor, 1999).

The 1000 kernel weight is an important attribute for analyzing the quality of the grain, its yield, and its further milling procedure (Shukla & Awasthi, 2017). Thilagavathi et al. (2015) reported 1000 grain weight of pearl millet (11.39 g/100 g) followed by kodo millet (2.45 g/100 g),

Table 3.1 Physical Properties of Millet Grains

Source	Bulk Density	1000 Kernel Weight (g)	1000 Seed Volume (ml)	Length (mm)	Breadth (mm)	References
Finger millet	993–1158 kg/ m³	496.8–775.8		1.41– 1.67	1.28–1.47	Ramashia et al. (2018)
		2.9	2.1			Shobana and Malleshi (2007)
Kodo millet	1.56 w/v	2.45	2.15			Thilagavathi et al. (2015)
Little millet	1.65 w/v	2.23	1.48			Thilagavathi et al. (2015)
Pearl millet	830.3– 866.1 kg/m³			2.98– 3.36	1.86–2.24	Jain and Bal (1997)
	1.75 w/v	11.39	6.4			Thilagavathi et al. (2015)
	0.830–0.877 g/ml					Siroha et al. (2016)
Proso millet	1.65 w/v	2.43	2.65			Thilagavahti et al. (2015)
		4.88–7.60		2.3–2.72	2.04–2.49	Yang et al. (2018)
	0.80–0.89 g/ ml	4.23–4.94	4.90–6.46	2–2.99	1.56–1.98	Bora and Das (2018)
	779.2					Devisetti et al. (2014)

proso millet (2.43 g/100 g), and little millet (2.23 g/100 g), respectively. The maximum seed volume was observed in pearl millet (6.46 ml) followed by proso millet (2.65 ml), kodo millet (2.15 ml), and little millet (1.48 ml). Moisture content is one of the important factors that govern the physical properties of grains (Goswami et al., 2015). It is also a good indicator as to whether the grains can be stored for a long or short period of time. Ramashia et al. (2018) studied the moisture content of the finger millet grain cultivar that ranged from 7.88% to 9.38%, while the moisture content of flours varied from 9.17% to 11.67%. According to Abdullah et al. (2012), the higher the moisture content, the shorter the storage life of the grain as high moisture content can cause rapid growth of mold on grains.

The mean results of the length, width, and thickness of the finger millet cultivars were 1.41–1.67, 1.28–1.47 and 1.22–1.35 mm, respectively (Ramashia et al., 2018). Similar results were obtained for pearl millet cultivars: length (2.98–3.36 mm), width (1.86–2.24 mm), and thickness (1.70–2.01 mm) (Jain & Bal, 1997). The geometric mean diameter ranged from 2.81 to 1.35 mm, and arithmetic mean diameter from 2.85 to 1.35 mm (Ramashia et al., 2018). The results for geometric and arithmetic mean diameters were similar to the results obtained on millets grains as reported by Adebowale et al. (2012b), where the average length, width, and thickness were 3.85, 2.06, and 2.05 mm. Jain and Bal (1997) studied the geometric and arithmetic mean diameters of pearl millet cultivars, and the results were 1.82–2.12 and 1.72–2.08 mm, respectively, at the moisture content of 7.4%. Porosity for millet grains was found to be 32.5%–63.7% (Jain & Bal, 1997; Balasubramanian &Viswanathan, 2010). Ramashia et al. (2018) also studied the porosity of finger millets, and the values varied from 24.31% to 32.41%. Aspect ratio and sphericity of various finger millet and pearl millet cultivars ranged from 73.55% to 92.21% and 73.75% to 94.25%, respectively (Jain & Bal, 1997; Ramashia et al. 2018).

The length, width, thickness, sphericity, and surface area of millet grains were found to be in the range of 3.31–4.21 mm, 1.82–2.27 mm, 1.90–2.33 mm, 0.60%–0.72%, and 16.92–24.38 mm^2, respectively, for Nigerian millet grains (Adebowale et al., 2012b). Bora and Das (2018) reported that the length, breadth, and thickness of whole proso millet grains were 2.99, 1.98, and 1.41 mm, respectively. It can be reported that milling reduced the grain sizes in terms of length, breadth, and thickness. Dehusking reduced the length (2.18 mm), breadth (1.86 mm), and thickness (1.04 mm) of proso millet grains. The length and breadth ratio of whole and dehusked proso millet grains were 1.51 and 1.17, respectively. The 1000 kernel weight of waxy and nonwaxy proso millet grains was between 4.88 and 7.60 g. The length and breadth of the grains were 2.36–2.72 and 2.04–2.49 mm, respectively. The length/breath of the grains ranged from 1.09 to 1.16, which indicated that the hulled proso millet grains were nearly spherical (Yang et al., 2018). Rao et al. (2019) studied physical properties of minor millets and observed different parameters: length (2.21–3.39 mm), width (1.87–2.78 mm), thickness (1.42–1.79 mm), aspect ratio (0.68–0.90), and 1000 kernel weight (1.83–4.38 g). Shobana and Malleshi (2007) reported the 1000 kernel weight (2.9 and 2.6 g) and 1000 kernel volume (2.1 and 1.7 ml) of native and decorticated finger millet grains.

3.3 MILLING PROPERTIES

Milling is an important and intermediate step in the postproduction processing of grains and in the transformation of raw materials into finer primary products for secondary processing (Bender, 2006). Figure 3.4 describes the milling methods of millets. Cereals are milled by grinding the grains using a roller, hammer, disk, or stone. With the grinding process, compression, impact, or shear force is applied to reduce the particle size. Different types of milling machines affect the physical, chemical, nutritional, and functional properties of the milled products (Aprodu & Banu, 2017).

3.4 FUNCTIONAL PROPERTIES

Functional properties of flours are those that primarily determine their utilization in different food products. The functional properties of the flours of different millets are shown in Table 3.2. Water absorption capacity (WAC) of millet flour was also observed as follows: pearl millet (153%–177%), proso millet (93.65%–111.50%), finger millet (0.93–1.23 ml/g), kodo millet (74.93 ml/100 g), and little millet (76.83 ml/100 g) (Thilagavathi et al., 2015; Siroha et al., 2016; Ramashia et al., 2018; Bora & Das, 2018).

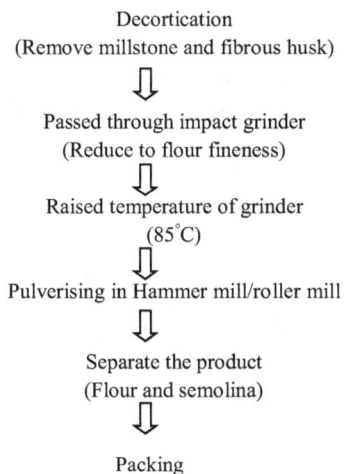

Decortication
(Remove millstone and fibrous husk)
⇩
Passed through impact grinder
(Reduce to flour fineness)
⇩
Raised temperature of grinder
(85°C)
⇩
Pulverising in Hammer mill/roller mill
⇩
Separate the product
(Flour and semolina)
⇩
Packing

Figure 3.4 Method of milling of millets.

Table 3.2 Functional Properties of Flours

Source	WAC	OAC	L*	a*	b*	References
Finger millet	0.93–1.23 ml/g		19.23–52.97	3.77–18.28		Ramashia et al. (2018)
Kodo millet	74.93 ml/100 g	74.74 ml/100 g	117.75	−6.25	12.40	Thilagavathi et al. (2015)
Little millet	76.83 ml/100 g	84.36 ml/100 g	121.23	−6.29	11.13	Thilagavathi et al. (2015)
Pearl millet	153%–177%	104%–124%	75.1–78.4	0.32–1.64	9.9–13.5	Siroha et al. (2016)
	74.08 ml/100 g	84.57 ml/100 g	103.94	3.26	37.97	Thilagavathi et al. (2015)
	1.33–1.42 g/g	0.94–1.25 g/g	46.51–47.21	4.81–6.18	19.43–22.18	Falade and Kolewole (2013)
Proso millet			64.78–71.23	2.04–3.57	24.73–27.08	Yang et al. (2018)
	93.65%–111.50%	100.67%–152.63%	56.37–63.04	0.79–4.18	18.46–25.17	Bora and Das (2018)
	75 ml/100 g	73 ml/100 g	126.62	−6.67	31.10	Thilagavathi et al. (2015)

The WAC of flour or isolate is a useful indicator for determining if the flour can be incorporated into aqueous food formulations, especially those involving dough handling. Lower WAC is suitable for making thinner gruels and also indicates the amount of water available for gelatinization (Giami, 1993). Adebowale et al. (2012a) mentioned that high WAC values indicate loose structure of starch polymers, while low values indicate the compactness of the structure.

Oil absorption capacity (OAC) was recorded to be 85.57 ml/100 g in pearl millet flour, and it was 84.36 ml/100 g in little millet flour, 74.74 ml/100 g in kodo millet flour, and 73.58 ml/100 g in proso millet flour (Thilagavathi et al., 2015) (Table 3.2). The type of proteins and the presence of non-polar side chains may also have an effect on OAC. OAC is the ability of a product to bind with oil (Di Cairano et al., 2020). The oil absorbing mechanism involves capillarity interaction, which allows the absorbed oil to be retained. Adebowale and Lawal (2004) reported that flours with good OAC are potentially useful in flavor retention, improvement of palatability, and extension of shelf life, particularly in bakery or meat products where fat absorption is desired.

Emulsion activity (EA) is the ability of flour to emulsify oil. Devisetti et al. (2014) observed the EA of proso millet flour to be 4.9 ml/100 ml, while that of pearl millet flour was observed to be 45%–60% (Siroha et al., 2016). Shrestha and Srivastava (2017) observed the EA of finger and barnyard millet flours as being 18.29% and 25.72%. The EA reflects the ability and capacity of a protein to aid in the formation of an emulsion and is related to the protein's ability to absorb oil and water in an emulsion (Du et al., 2014). The efficiency of emulsification varies with the type, concentration, and solubility of the proteins (Achinewhu, 1983).

The foaming capacity of pearl millet flours ranged from 18% to 25% (Siroha et al., 2016). Foaming capacity of proso millet flour varied from 0.20 to 8.7 (ml/100 ml) (Devisetti et al., 2014; Kumar et al., 2020). Foaming capacity is an important functional characteristic of flours that determines their utilization in food systems, where aeration and overrun are required (Shevkani et al., 2014). Owing to a large increase in the surface area in the liquid/air interphase, proteins denature and aggregate during whipping. This property is important for flour used in many leavening food products such as baked goods, cakes, and biscuits (Belitz & Grosch, 1999). Foam formation is governed by three factors: transportation, penetration, and reorganization of the molecule at the air–water interface. Therefore, for good foaming, the protein should be capable of migrating at the air–water interface, unfolding, and rearranging at the interface (Sreerama et al., 2008).

3.5 COLOR PROPERTIES

Color is an important sensory characteristic and is used to judge the quality and acceptance of food or products (Clydesdale, 1993). Hunter color parameters of cereal grains are evaluated using the hunter color lab. Varietal differences are observed for various color parameters of millets. The L* value indicates the lightness, with 0–100 representing dark to light, and values ranging from 89.2 to 92.7. The measurement of color (L*, a*, and b*) values of different millet flours (Kodo, little, pearl and proso millet) were found to be 103.94–126.62, -3.26- -6.67, and 11.13–37.97, respectively (Thilagavathi et al., 2015). The highest b* value was reported for pearl millet, which shows a more yellowish color compared with other millets (Thilagavathi et al., 2015). The L*, a*, and b* values of finger millets were found to be 31.1, 11.8, and 13.6, respectively (Onyango et al., 2020). Yang et al. (2018) studied the color properties the proso millet grains and found that L* value of nonwaxy proso millet grains ranged from 64.78 to 67.89 while that for waxy proso millets ranged from 69.26 to 71.23. The total color difference of nonwaxy proso millet ranged from 38.63 to 40.29, and the total color difference of waxy proso millet ranged from 35.94 to 37.61. Yadav et al. (2012) reported hunter L*, a*, and b* values of 79.3, 1.23, and 12.6, respectively, for pearl millet flour, while Falade and Kolawole (2013) observed the color properties of two pearl millet cultivars and observed values of L*, a*, and b* of 46.51–47.21, 4.81–6.18, and 19.43–22.18, respectively.

3.6 PASTING PROPERTIES

The pasting property represents changes in the behavior of flour paste viscosity with change in temperature; this mainly varies with flour composition and the characteristics of starch (Bao & Bergman, 2004). The pasting process is usually studied with an amylograph, a Rapid Visco Analyzer (RVA), a Brabender Amylograph, an Ottawa Starch Viscometer, or a dynamic Rheometer. These instruments record the change in viscosity of starch pastes/suspensions under shear force as a function of temperature (Punia et al., 2020). The pasting profiles of a starch are an effective method for relating starch functionality with its structural features and enables identifying the suitable product for the correct industrial application, especially for those products dependent on the viscosity and thickening behavior of starch (Cruz et al., 2013). During pasting property measurement, flour showed increase in the viscosity with increase in temperature, a drop in viscosity when temperature was kept constant, followed by an increase in the viscosity again at the end of cooling cycle (Figure 3.5). Peak viscosity (PV) of millet flour varied from 652 to 2262 cP, with the highest value observed for barnyard millet flour (Yadav et al., 2012; Sun et al., 2014; Goswami et al., 2015; Yang et al., 2018). Peak viscosity (PV) provides information for the maximum swelling of the starch granule prior to disintegration (Liu et al., 2006). The breakdown viscosity (BV) is a measurement of degree of disintegration of granules and is measured as the difference between the PV and the trough viscosity (TV). The BV of flours ranged between 186 and 1008 cP (Yadav et al., 2012; Sun et al.,

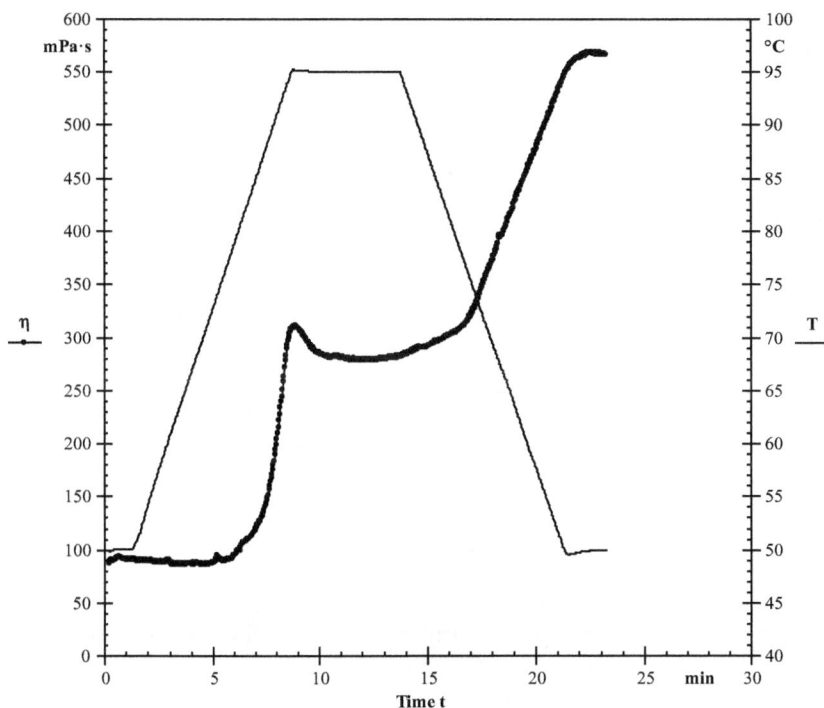

Figure 3.5   Pasting graph of millet flour.

2014; Yang et al., 2018). The breakdown is caused by the disintegration of the gelatinized starch granule structure during continuous stirring and heating (Whistler & BeMiller, 1997). According to Lee et al. (1995), breakdown viscosity is influenced by amylose content. Goswami et al. (2015) studied the effect of millet flour addition in wheat flour on pasting properties and observed a decrease in PV with further addition of millet flour. This may be due to the increase in fiber, total minerals, and lipid content with increasing level of the millet flour that competes with starch granules for water absorption, thus obstructing the swelling of starch granules (Nimsung et al., 2007). The setback viscosity (SV) of starch is the difference between the final viscosity and trough viscosity. SV of millet ranges between 153 and 3644 cP (Yadav et al., 2012; Sun et al., 2014; Goswami et al., 2015; Yang et al., 2018). Final viscosities (FVs) of proso millet (261–4705 cP), pearl millet (68.11 BU & 1938 cP), and barnyard millet (53.83–138.08 RVU) were also assessed (Kim et al., 2011; Sun et al., 2014; Akinola et al., 2017; Yang et al., 2018). The FV of starches is influenced by the aggregation of amylose molecules (Blazek & Copeland, 2008), whereas SV is the measure of the gelling ability or the retrogradation tendency of starch (Shelton & Lee, 2000). Pasting temperature (PT) is the minimum temperature required to cook the flour. PT of millet flours varied from 67.15°C to 90.3°C, with the highest PT observed for barnyard millet flour (Kim et al., 2011; Sun et al., 2014; Akinola et al., 2017; Yang et al., 2018; Kumar et al., 2020). The pasting temperature can probably be affected by factors such as the degree of branching of amylopectin and a higher degree of crystallinity (Kim et al., 1996). It is difficult to directly compare pasting results of millet flours with other commercial flours like wheat, rice, and corn flours due to the need to use different instruments and experimental conditions for measuring pasting properties in different flours (Table 3.3).

3.7 THERMAL PROPERTIES

The cooking quality of flours is measured by gelatinization of flours, which has a major role in processing of food. Gelatinization temperatures are also important for selecting specific properties of starches according to requirements in various food applications (Morales-Martínez et al., 2014). The gelatinization can be quantified by various methods, with DSC being the most commonly used method for millet flours analysis where onset (T_o), peak (T_p), and conclusion temperatures (T_c), and enthalpy change (ΔH) are measured (Table 3.4). To obtain consistent results of starch gelatinization temperature and ΔH, the sample must contain at least two times (w/w) of water. Without a sufficient amount of water, the gelatinization peak broadens and shifts to a higher temperature (Donovan, 1979). Shinoj et al. (2006) studied thermal properties (onset T_o, peak T_p, and endset or conclusion T_c temperatures and gelatinization enthalpies H) of minor millets (finger millet, proso millet, foxtail millet, kodo millet, little millet, barnyard millet). Gelatinization temperatures T_o, T_p, and T_c varied from 71.2°C to 76.8°C, 73.1°C to 78.7°C, and 87.7°C to 101.8°C, respectively, with the highest gelatinization temperature being observed for proso millet flour. High transition temperatures are believed to result from high degree of crystallinity, which imparts structural stability, making the granules more resistant to gelatinization (Barichello et al., 1990). Enthalpy of gelatinization was found to be 5.73–10.33 KJ/kg. Kumar et al. (2020) and Sun et al.

Table 3.3 Pasting Properties of Flours

Source	Peak Viscosity	Breakdown Viscosity	Setback Viscosity	Final Viscosity	Unit	Pasting Temperature (°C)	References
Barnyard millet	38.7–98.75			53.83–138.08	RVU	83.2–90.3	Kim et al. (2011)
	2262		2421	4136	cP		Goswami et al. (2015)
Finger millet	1566	298	2422.6	3684	cP		Kharat et al. (2019)
Foxtail millet	2392	540	3396.3	578.6	cP		Kharat et al. (2019)
Pearl millet	1362	649	1226	1938	cP		Yadav et al. (2012)
	56.56	17.52	29.04	68.11	BU	68.0	Akinola et al. (2017)
	1950	420		4433.3	cP		Kharat et al. (2019)
Proso millet	652	543	153	261	cP	75.2	Sun et al. (2014)
	891–2116	186–1008	197–3644	901–4705	cP	78–87	Yang et al. (2018)
	2153.64	21.63	4.75	2125.83	RVU	67.15	Kumar et al. (2020)

Table 3.4 Thermal Properties of Millet Flours

Source	Flour: Water Ratio (w/w)	Heating Rate (°C/min)	Gelatinization Parameters				
			T_o (°C)	T_p (°C)	T_c (°C)	ΔH gel	References
Barnyard millet	1:3	10	74.9	76.7	87.7	6.44 KJ/ Kg	Shinoj et al. (2006)
Finger millet	1:5	5	71.52	76.71	86.57	15.05 J/g	Kharat et al. (2019)
	1:3	10	71.2	73.1	88.2	10.33 KJ/Kg	Shinoj et al. (2006)
Foxtail millet	1:5	5	73.52	78.01	91.40	6.95 J/g	Kharat et al. (2019)
	1:3	10	73.4	75.9	101.8	10.15 KJ/Kg	Shinoj et al. (2006)
Kodo millet	1:3	10	76.6	78.4	90.0	6.76 KJ/ Kg	Shinoj et al. (2006)
Little millet	1:3	10	75.8	77.5	87.8	5.73 KJ/ Kg	Shinoj et al. (2006)
	1:3	10	71.26	78.44	82.28	1.69 J/g	Kumar et al. (2020)
Pearl millet	1:5	5	71.35	76.15	85.22	4.32 J/g	Kharat et al. (2019)
Proso millet	1:3	10	76.8	78.7	89.4	7.89 J/g	Shinoj et al. (2006)
	1:2	10	67.62	70.15	83.56	15.03 J/g	Sun et al. (2014)
	1:3	10	64.56	72.68	80.09	0.98 J/g	Kumar et al. (2020)

(2014) reported T_o, T_p, and T_c values of 64.56°C–67.62°C, 70.15°C–72.68°C, and 80.09°C–83.56°C for proso millet flour, respectively. The thermal properties of finger millets are as follows: 68.16°C, 70.62°C, 73.79°C, and 9.64 J/g for T_o, T_p, T_c, and ΔH, respectively (Adebowale et al., 2005). Kharat et al. (2019) conduct a comparative study on thermal properties of finger, foxtail, and pearl millet flours. T_o, T_p, and T_c values of finger millet (71.52°C, 70.15°C, 83.56°C), foxtail millet (73.52°C, 78.01°C, 91.40°C), and pearl millet (71.35°C, 76.15°C, 85.22°C) were obtained as well. It was observed that pearl millet showed the highest gelatinization. The diversity in gelatinization may be attributed to the molecular structure and content of amylopectin and amylose, the presence of lipids, and the physical association and orientation of these chemical components in the granules (Srichuwong & Jane, 2007). ΔH gel is an indicator of the loss of molecular double-helical order within the granule (Cooke & Gidley, 1992; Tester & Morrison, 1990). The lower ΔH gel suggests a lower percentage of organized arrangements or a lower stability of the crystals (Chiotelli & Meste, 2002).

3.8 IMPORTANCE OF PHYSICAL AND FUNCTIONAL PROPERTIES

Quality of grains depends on two different factors: (1) intrinsic factors and (2) extrinsic factors. Intrinsic factors include such ones as bulk density, color, aroma, shape, size, and nutritional composition. Color is considered as one of the important characteristic features of grains as it helps in grading during processing. Color is also an indication of quality, which sometimes varies with cultivar type, whereas sometimes it may indicate the types of infection and deficiencies present in the grains. Bulk density demonstrates the ratio of the mass to a given volume of a grain sample, which also includes the interstitial voids between particles. The test weight of grains sample is the traditional and simplest way to estimate the grains quality. In general, grain quality depends on their test weight, which has a direct relationship with their quality attributes. Furthermore, shape and size are two important factors that help in quality judgment and grading. However, these two features may vary between grain types and cultivars (Bern & Brumm, 2009; Ram et al., 2002; Henry & Kettlewell, 1996).

Extrinsic factors include age of grain, damages in grains, immaturity, presence of foreign materials, type of infection, and moisture content in specific grains and cultivars. Before processing of grains at the industrial level, it is important to make sure that they are soft enough for processing as harder grains makes it processing them into flour during milling difficult. Hard grains take more time to get converted to the flour form. Hardness is the desirable feature that millers as well as traders use to classify them into subcategories (Serna-Saldivar, 2012).

Millet grains are a good source of starch, protein, fibers, antioxidants, and minerals. In-depth information on functional properties helps to estimate the behavior of flour under different processing conditions.

Functional properties of flour describe how its nutrient components such as starch, fiber, proteins, minerals, fat, and bioactive compounds behave during processing and how much they could affect the quality of the finished food product in terms of look, taste, and feel. The functional properties determine whether the blends of different flour types would be useful in bakery products where hydration is carried out to improve texture and other desirable features. Millets are extensively being studied and used in the production of various food products. The functional properties of millet flour are an essential parameter to describe quality in terms of appearance, organoleptic properties and acceptance by consumers.

3.9 CONCLUSION

Physical properties are indicators of the quality of cereal grains. For designing the equipment for various the various processing techniques for millets, physical characteristics are very important. Physical properties of grains include 1000 kernel weight, moisture content, porosity, bulk density, sample volume, aspect ratio, true density, sample surface area, and perpendicular dimensions (length, width, and thickness). The physical behavior, sensory characteristics, and end quality of food products are associated with physical, physicochemical, and functional properties of millets. The functional properties are the properties of flours that primarily determine their utilization in different food products. The use of grain flours in food formulations is dependent on flour functionality. The water absorption capacity, oil absorption capacity, emulsion capacity, and foaming capacity play important roles in the development of food products.

REFERENCES

Abdullah, M. H. R. O., Ch'ng, P. E. and Yunus, N. A. 2012. Some physical properties of musk lime (*Citrus Microcarpa*). *International Journal of Biological, Biomolecular, Agricultural, Food and Biotechnological Engineering* 6: 1122–1125.

Achinewhu, S. C. 1983. Protein and food potential of African oil bean (*Pentaclethra macrophylla*) and velvet bean (*Mucunauries*). *Journal of Food Science* 47: 1736–1742.

Adebowale, A. A., Adegoke, M. T., Sanni, S. A., Adegunwa, M. O. and Fetuga, G. O. 2012a. Functional properties and biscuit making potentials of sorghum-wheat flour composite. *American Journal of Food Technology* 7: 372–379.

Adebowale, K. O., Afolabi, T. A. and Olu-Owolabi, B. I. 2005. Hydrothermal treatments of Finger millet (*Eleusine coracana*) starch. *Food Hydrocolloids* 19: 974–983.

Adebowale, A. A., Fetuga, G. O., Apata, C. B. and Sannai, L. O. 2012b. Effect of variety and initial moisture content on physical properties of improved millet grains. *Nigerian Food Journal* 30: 5–10.

Adebowale, K. O. and Lawal, O. S. 2004. Comparative study of the functional properties of bambarra groundnut (*Voandzeia subterranean*), jack bean (*Canavalia ensiformis*) and mucuna bean (*Mucuna pruriens*) flours. *Food Research International* 37: 355–365.

Akinola, S. A., Badejo, A. A., Osundahunsi, O. F. and Edema, M. O. 2017. Effect of preprocessing techniques on pearl millet flour and changes in technological properties. *International Journal of Food Science & Technology* 52: 992–999.

Akpata, M. I. and Akubor, P. I. 1999. Chemical composition and selected functional properties of sweet orange (*Citrus sinensis*) seed flour. *Plant Foods for Human Nutrition* 54: 353–362.

Aprodu, I. and Banu, I. 2017. Milling, functional and thermo-mechanical properties of wheat, rye, triticale, barley and oat. *Journal of Cereal Science* 77: 42–48.

Asoegwu, S. N., Ohanyere, S. O., Kanu, O. P. and Iwueko, C. N. 2006. Physical properties of African oil beans (*Pentaclethra macroplaylla*). *Agricultural Engineering International: The IGR E Journal*, 8: 1–16.

Balasubramanian, S. and Viswanathan, R. 2010. Influence of moisture content on physical properties of minor millets. *Journal of Food Science & Technology* 47: 279–284.

Bao, J., Bergman, C. J. 2004. The functionality of rice starch. In: Eliasson, A. C., editor, *Starch in food: structure, function and applications*. Cambridge: Woodhead Publishing Ltd. pp. 258–294.

Barichello, V., Yada, R. Y., Coffin, R. H. and Stanley, D. W. 1990. Low temperature sweetening in susceptible and resistant potatoes: starch structure and composition. *Journal of Food Science* 55: 1054–1057.

Baryeh, E. A. 2001. Physical properties of bambara groundnuts. *Journal of Food Engineering* 47(4): 321–326.

Belitz, H. D. and Grosch, W. 1999. *Food chemistry*. 2nd ed. Berlin: Springer.

Bender, D. A. 2006. *Benders Dictionary of Nutrition and Food Technology*. 8th ed. Abington: Woodhead Publishing/CRC Press.

Bern, C. and Brumm, T. J. 2009. Grain test weight deception. Iowa State University-University Extension. PMR 1005, October 2009.

Blazek, J. and Copeland, L. 2008. Pasting and swelling properties of wheat flour and starch in relation to amylose content. *Carbohydrate Polymers* 71: 380–387.

Bora, P. and Das, P. 2018. Some physical and functional properties of proso millet (*Panicum miliaceum* L.) grown in Assam. *International Journal of Pure & Applied Bioscience* 6: 1188–1194.

Chandra, S. and Singh, S. 2013. Assessment of functional properties of different flours. *African Journal of Agricultural Research* 8: 4849–4852.

Chiotelli, E. and Meste, M. L. 2002. Effect of small and large wheat starch granules on thermomechanical behaviour of starch. *Cereal Chemistry* 79: 286–293.

Clydesdale, F. M. 1993. Color as a factor in food choice. *Critical Reviews in Food Science & Nutrition* 33(1): 83–101.

Cooke, D. and Gidley, M. J. 1992. Loss of crystalline and molecular order during starch gelatinization: origin of the enthalpic transition. *Carbohydrate Research* 227: 103–112.

Cruz, B. R., Abraão, A. S., Lemos, A. M. and Nunes, F. M. 2013. Chemical composition and functional properties of native chestnut starch (*Castanea sativa Mill*). *Carbohydrate Polymers* 94: 594–602.

Davies, R. M. 2009. Some physical properties of groundnut grains. *Research Journal of Applied Sciences, Engineering & Technology* 1(2): 10–13.

Devisetti, R., Yadahally, S. N. and Bhattacharya, S. 2014. Nutrients and antinutrients in foxtail and proso millet milled fractions: evaluation of their flour functionality. *LWT-Food Science & Technology* 59: 889–895.

Di Cairano, M., Condelli, N., Caruso, M. C., Marti, A., Cela, N. and Galgano, F. 2020. Functional properties and predicted glycemic index of gluten free cereal, pseudocereal and legume flours. *LWT-Food Science & Technology* 133: 109860.

Donovan, J. W. 1979. Phase-transitions of the starch-water system. *Biopolymers* 18: 263–275.

Du, S. K., Jiang, H., Yu, X. and Jane, J. L. 2014. Physicochemical and functional properties of whole legume flour. *LWT-Food Science & Technology* 55: 308–313.

Esref, I. and Halil, Ü. 2007. Moisture-dependent physical properties of white speckled red kidney bean grains. *Journal of Food Engineering* 82: 209–216.

Falade, K. O. and Kolawole, T. A. 2013. Effect of γ-irradiation on colour, functional and physicochemical properties of pearl millet [*Pennisetum glaucum* (L) R. Br.] cultivars. *Food & Bioprocess Technology* 6(9): 2429–2438.

Faleye, T., Atere, A. O., Oladipo, O. N. and Agaja, M. O. 2013. Determination of some physical and mechanical properties of some cowpea varieties. *African Journal of Agricultural Research* 8: 6485–6487.

Giami, S. Y. 1993. Effect of processing on the proximate composition and functional properties of cowpea (*Vigna unquiculata*) flour. *Food Chemistry* 47: 153–158.

Goswami, D., Gupta, R. K., Mridula, D., Sharma, M. and Tyagi, S. K. 2015. Barnyard millet based muffins: physical, textural and sensory properties. *LWT-Food Science and Technology* 64: 374–380.

Henry, R. J. and Kettlewell, P. S. 1996. *Cereal grain quality*. 1st ed. London, UK: Chapman and Hall.

Jain, R. K. and Bal, S. 1997. Properties of pearl millet. *Journal of Agricultural Engineering & Research* 66: 85–91.

Kaushal, P., Kumar, V. and Sharma, H. K. 2012. Comparative study of physicochemical, functional, antinutritional and pasting properties of taro (*Colocasia esculenta*), rice (*Oryza sativa*) flour, pigeonpea (*Cajanus cajan*) flour and their blends. *LWT-Food Science & Technology* 48: 59–68.

Kharat, S., Medina-Meza, I. G., Kowalski, R. J., Hosamani, A., Ramachandra, C. T., Hiregoudar, S. and Ganjyal, G. M. 2019. Extrusion processing characteristics of whole grain flours of select major millets (foxtail, finger, and pearl). *Food & Bioproducts Processing* 114: 60–71.

Kheiralipour, K., Karimi, M., Tabatabaeefar, A., Naderi, M., Khoubakht, G. and Heidarbeigi, K. 2008. Moisture-depend physical properties of wheat (*Triticum aestivum* L.). *Journal of Agricultural Technology* 4: 53–64.

Kim, J. Y., Jang, K. C., Park, B. R., Han, S. I., Choi, K. J., Kim, S. Y., Oh, S. H., Ra, J. E., Ha, T. J., Lee, J. H. and Hwang, J. 2011. Physicochemical and antioxidative properties of selected barnyard millet (*Echinochloa utilis*) species in Korea. *Food Science & Biotechnology* 20: 461–469.

Kim, Y. S., Wiesenborn, D. P., Lorenzen, J. H. and Berglund, P. 1996. Suitability of edible bean and potato starches for starch noodles. *Cereal Chemistry* 73: 302–308.

Kumar, S. R., Sadiq, M. B. and Anal, A. K. 2020. Comparative study of physico-chemical and functional properties of pan and microwave cooked under-utilized millets (proso and little). *LWT-Food Science & Technology* 128: 109465.

Kumari, N. and Raghuvanshi, R. S. 2015. Physico-chemical and functional properties of buckwheat (*Fagopyrum esculentum* Moench). *Journal of Eco-friendly Agriculture* 10: 77–81.

Lee, M. H., Hettiarachchy, N. S., McNew, R. W. and Gnanasambandam, R. 1995. Physicochemical properties of calcium-fortified rice. *Cereal Chemistry* 72: 352–355.

Liu, Q., Donner, E., Yin, Y., Huang, R. L. and Fan, M. Z. 2006. The physicochemical properties and in vitro digestibility of selected cereals, tubers and legumes grown in China. *Food Chemistry* 99: 470–477.

Morales-Martínez, L. E., Bello-Pérez, L. A., Sánchez-Rivera, M. M., Ventura-Zapata, E. and Jiménez-Aparicio, A. R. 2014. Morphometric, physicochemical, thermal, and rheological properties of rice (*Oryzasativa* L.). *Food & Nutrition Sciences* 5: 271–279.

Nimsung, P., Thongngam, M. and Naivikul, O. 2007. Compositions, morphological and thermal properties of green banana flour and starch. *Kasetsart Journal Natural Science* 41: 324–330.

Onyango, C., Luvitaa, S. K., Unbehend, G. and Haase, N. 2020. Nutrient composition, sensory attributes and starch digestibility of cassava porridge modified with hydrothermally-treated finger millet. *Journal of Agriculture & Food Research* 2: 100021.

Punia, S., Sandhu, K. S., Dhull, S. G., Siroha, A. K., Purewal, S. S., Kaur, M. and Kidwai, M. K. 2020. Oat starch: physico-chemical, morphological, rheological characteristics and its application–a review. *International Journal of Biological Macromolecules* 154: 493–498.

Ram, M. S., Dowell, F. E., Seitz, L. and Lookhart, G. 2002. Development of standard procedures for a simple, rapid test to determine wheat color class. *Cereal Chemistry* 79(2): 230–237.

Ramashia, S. E., Gwata, E. T., Meddows-Taylor, S., Anyasi, T. A. and Jideani, A. I. O. 2018. Some physical and functional properties of finger millet (*Eleusine coracana*) obtained in sub-Saharan Africa. *Food Research International* 104: 110–118.

Rao, B. D., Sharma, S., Kiranmai, E. and Tonapi, V. A. 2019. Effect of processing on the physico-chemical parameters of minor millet grains. *International Journal of Chemical Studies* 7: 276–281.

Serna-Saldivar, S. O. 2012. *Cereal grains laboratory reference and procedures manual.* Food Preservation Technology Series. LLC NW, Boca Raton: CRC Press. Taylor and Francis Group.

Shelton, D. R. and Lee, W. J. 2000. Food science and technology. In: Kulp, K., Ponte, J. G., editors. *Cereal carbohydrates*. Boca Raton, FL: CRC Press. pp. 385–416.

Shevkani, K., Singh, N., Kaur, A. and Rana, J. C. 2014. Physicochemical, pasting, and functional properties of amaranth seed flours: effects of lipids removal. *Journal of Food Science* 79: 1271–1277.

Shinoj, S., Viswanathan, R., Sajeev, M. S. and Moorthy, S. N. 2006. Gelatinisation and rheological characteristics of minor millet flours. *Biosystems Engineering* 95: 51–59.

Shobana, S. and Malleshi, N. G. 2007. Preparation and functional properties of decorticated finger millet (*Eleusine coracana*). *Journal of Food Engineering* 79: 529–538.

Shrestha, R. and Srivastava, S. 2017. Functional properties of finger millet and banyard millet flours and flour blends. *International Journal of Science & Research* 6: 775–780.

Shukla, S. and Awasthi, P. 2017. Physical properties of green gram and tamarind kernel and analysis of functional properties of composite flours incorporating tamarind kernel powder. *International Journal of Science & Research* 6: 632–637.

Siddiq, M., Nasir, M., Ravi, R., Dolan, K. D. and Butt, M. S. 2009. Effect of defatted maize germ addition on the functional and textural properties of wheat flour. *International Journal of Food Properties* 12: 860–870.

Siroha, A. K., Sandhu, K. S. and Kaur, M. 2016. Physicochemical, functional and antioxidant properties of flour from pearl millet varieties grown in India. *Journal of Food Measurement & Characterization* 10: 311–318.

Sreerama, Y. N., Sasikala, V. B. and Pratape, V. M. 2008. Nutritional implications and flour functionality of popped/expanded horse gram. *Food Chemistry* 108: 891–899.

Srichuwong, S. and Jane, J. L. 2007. Physicochemical properties of starch affected by molecular composition and structures: a review. *Food Science & Biotechnology* 16: 663–674.

Subramanian, S. and Viswanathan, R. 2003. Thermal properties of minor millet grains and flours. *Biosystems Engineering* 84: 289–296.

Subramanian, S. and Viswanathan, R. 2007. Bulk density and friction coefficients of selected minor millet grains and flours. *Journal of Food Engineering* 81: 118–126.

Sun, Q., Gong, M., Li, Y. and Xiong, L. 2014. Effect of dry heat treatment on the physicochemical properties and structure of proso millet flour and starch. *Carbohydrate Polymers* 110: 128–134.

Tester, R. F. and Morrison, W. R. 1990. Swelling and gelatinization of cereal starches. *Cereal Chemistry* 67: 558–563.

Teunou, E. and Fitzpatrick, J. J. 1999. Effect of relative humidity and temperature on food powder flowability. *Journal of Food Engineering* 42: 109–116.

Thilagavathi, T., Kanchana, S., Banumathi, P., Hemalatha, G., Vanniarajan, C., Sundar, M. and Ilamaran, M. 2015. Physico-chemical and functional characteristics of selected millets and pulses. *Indian Journal of Science & Technology* 8: 147–155.

Vadivoo, A. S., Joseph, R. and Ganesan, N. M. 1998. Genetic variability and diversity for protein and calcium contents in finger millet (*Eleucine coracona* L) Gaertn in relation to grain color. *Plant Foods for Human Nutrition* 52: 353–364.

Varnamkhasti, M. G., Mobli, H., Jafari, A., Keyhani, A. R., Soltanabadi, M. H., Rafiee, S. and Kheiralipour, K. 2008. Some physical properties of rough rice (*Oryza Sativa* L.) grain. *Journal of Cereal Science* 47: 496–501.

Whistler, R. L. and BeMiller, J. N. 1997. *Carbohydrate chemistry for food scientist*. St. Paul, MN: Amercian Association of Cereal Chemists. pp. 117–151.

Yadav, D. N., Kaur, J., Anand, T. and Singh, A. K. 2012. Storage stability and pasting properties of hydrothermally treated pearl millet flour. *International Journal of Food Science & Technology* 47: 2532–2537.

Yang, Q., Zhang, P., Qu, Y., Gao, X., Liang, J., Yang, P. and Feng, B. 2018. Comparison of physicochemical properties and cooking edibility of waxy and non-waxy proso millet (*Panicum miliaceum* L.). *Food chemistry* 15: 271–278.

4

Nutritional Composition and Health Benefits

4.1 INTRODUCTION

Millets are important cereal grains characterized by small seeds with the capability to grow even under harsh environmental conditions. Throughout the world, millets are being utilized as an important source of food and feed (Salar & Purewal, 2017; Siroha et al., 2016). They are considered as one of the important crops in African and Asian regions, especially India, Nigeria, Mali, and Niger. Millets are favored all over the world because of their productivity and short maturation span. The crop can help maintain sustainability in the agro-industrial sector. Millets have the potential to grow in conditions of minimal water/drought, and they could also be considered as a water-saving crop as they pose no major risk to ground water level (Izumi et al., 2018; Rockström & de-Rouw, 1997). Millets are fulfilling the rising demands for food and provide the required nutrients that are necessary to maintain biochemical reaction mechanisms in living organisms. Millet grains are good sources of dietary as well as health-benefit components such as carbohydrates, proteins, fiber, minerals/vitamins, and bioactive compounds (Purewal et al., 2020; Salar et al., 2017a; Saleh et al., 2013; Devi et al., 2011). The gluten-free nature of millet grains makes them more popular among consumers as compared to gluten-rich natural substrates (Taylor & Emmambux, 2008; Brasil et al., 2015). Millets could, thus, be a better option for those who are in need of gluten-free food products. Analysis of nutritional composition is important from the industrial and health points of view. Nutritional analysis helps industrialists prepare specific food products per the consumer demand. Further, specific functional properties of grains help in predetermination of cooking quality and other important parameters. The behavior of flour and its paste under a specific set of conditions helps to prepare products accordingly (Siroha et al., 2020; Sandhu & Siroha, 2017). Grains may have many nutrients in them, some of which are macronutrients, while others are micronutrients. The presence of bioactive constituents in grains makes them as important as medicines as they play a vital role in maintaining healthy life (Purewal et al., 2019; Salar et al., 2017b). On the basis of nutritional profile, it becomes easier to prepare dietary chart for daily requirements for people. In-depth information regarding health benefits helps in eradicating disease conditions. Nutritional information helps the consumers choose the products as per their needs. This chapter provides in-depth information about the nutritional composition and health benefits of millets.

4.2 NUTRITIONAL COMPOSITION

Determination of nutritional composition is a key factor that can solve the food insecurity problems and may help to overcome malnutrition (Singh & Raghuvanshi, 2012). Millet grains are good sources of many health-benefiting nutrient components. The major proportion in the grains is occupied by starch followed by protein, fat, fibers, minerals/vitamins, and bioactive compounds. The approximate composition of various components in millets is as follows: crude protein (7%–15%), crude fat (1%–7%), crude fiber (2%–14%), and ash (2%–5%).

4.2.1 Carbohydrates

Milled millet grains possess different proportions of nutrients such as starch, ranging from 60% to 75%, and nonstarchy polysaccharides (15%–20%) and sugars (2%–3%) (Himashu et al., 2018). The sugars profile of millets indicates the presence of sucrose (31%–35%), glucose (9.9%–15%), fructose (8.6%–15%), raffinose (8.6%–12%), maltose (9%–11%), maltotriose (5%–6.1%), and xylose (1.5%–4.3%) in significant proportions (Wankhede et al., 1979). The nonstarchy part includes dietary fibers such as hemicelluloses, cellulose, and the pectinacious part. Starch is one of the industrially important nutrients that form the major carbohydrate proportion in millet grains. Rheological and pasting properties of specific millet starches illustrate their behavior under a specific set of conditions, ultimately demonstrating their use in different food products (Punia et al., 2020, 2019; Siroha et al., 2019). Starch could be used in the preparation of various food products as well as biodegradable packing materials (Sandhu et al., 2020). Kumari and Thayumanavan (1998) studied starch present in different millet types and observed that characteristic shape and size of starch granules vary with the type of millets. The shape of proso millet starch was small spherical, large spherical, or (rare) large polygonal, with specific sizes ranging from 1.3 to 8.0 μm. Foxtail millet possesses small spherical, small polygonal, and large pentagonal shapes with sizes of 0.8–9.6 μm. In barnyard millet, the shape of starch granules is small spherical, large spherical, and large polygonal, with sizes ranging from 1.2 to 10 μm. In kodo millet, the shape of starch granules was large polygonal, rarely small spherical, and polygonal, with sizes of 1.2–9.5 μm. In little millet, the shape of the starch granules observed was small spherical and large spherical, with a size range of 1.0–9.0 μm. The amount of amylose content in millets may vary with the type. For instance, the amount of amylose content reported for pearl millet was 13.6–18.1 g/100 g followed by finger millet (30–32.4 g/100 g), proso millet (1.2–21.5 g/100 g), barnyard millet (18–31 g/100 g), foxtail millet (3.3–25.21 g/100 g), and kodo millet (18–19.61 g/100 g) (Bean et al., 2019; Sandhu & Siroha, 2017; Annor et al., 2014; Wu et al., 2014; Balasubramanian et al., 2011; Liu et al., 2010; Kim et al., 2009; Kumari & Thayumanavan, 1998).

4.2.2 Protein

The nutritional profile of millets indicates the presence of protein in millets in the following order: pearl millet (4%–21%), foxtail millet (9%–19%), proso millet (12%–16%), finger millet (5%–14%), and kodo millet (6%–13%) (Boyles et al., 2017; Rhodes et al., 2017; Sukumaran et al., 2012; Singh & Raghuavanshi, 2012; Nambiar et al., 2011; Murray et al., 2008; Hooks et al., 2006; Waniska & Rooney, 2000). The specific fraction of proteins includes prolamin, glutelin, globulin, and albumin. The content of glutelin fractions and prolamin dominates over other protein types such as albumins and globulins. The amount of protein may vary with cultivar type and type of millets. For instance, of the total protein content, the amount of prolamin was observed to be high in foxtail millet (47.6%–63.4%), finger millet (24.6%–36%), and pearl millet (23%–31%) (Himashu et al., 2018; Geervani & Eggum, 1989; Singh et al., 1987; Indira & Naik, 1971). Glutelin was also observed in different millets in different percentages: kodo millet (40%–52%), pearl millet (13%–30%), finger millet (12%–28%), and foxtail millet (5%–18%) (Cremer et al., 2014; Adebowale et al., 2011; Kumar & Parameswaran, 1998; Serna-Saldivar & Rooney, 1995; Vivas et al., 1992; Subramanian et al., 1990). Analytical reports on millet proteins demonstrated that globulin and albumin are present mainly in the germ portion and outer layer of millet grains. For the exact estimation of protein structure and functionality, it is necessary to predict the presence of specific amino acids in them. The type and amount of amino acids may vary in different types of millets. The amino acid profile demonstrated that pearl millet is rich in leucine (80–251 g/Kg), lysine (17–65 g/Kg), tryptophan (11–28 g/Kg), aspartic acid (49–103 g/Kg), and proline (59–142 g/Kg), whereas foxtail millet protein is rich in arginine (27–95 g/Kg). Similarly finger millet is rich in phenylalanine (32–84 g/Kg), isoleucine (37–85 g/Kg), methionine (13–43 g/Kg), threonine (27–58 g/Kg), valine (58–104 g/Kg), glutamic acid (203–378 g/Kg), cysteine (7–29 g/Kg), glycine (33–59 g/Kg), serine (51–87 g/Kg), and tyrosine (20–56 g/Kg). Protein isolated from proso millet is rich in histidine (18–29 g/Kg) and alanine (39–122 g/Kg) (Table 4.1).

4.2.3 Lipids

Lipids are also important from the nutritional point of view as they help in various reactions in the biological system. Millet grains have fat content in range of 1%–5%, with their maximum distribution in the bran and endosperm (Himashu et al., 2018). A major proportion of fat (24%) is present in millet grains embryo. The fatty acid profile of millets indicates the presence of both types of fatty acids (saturated and unsaturated). The maximum proportion in millet grains is made of unsaturated fatty acid, ranging from 78% to 82%; saturated fatty acids account for 17.9%–21.6%. Osagie and Kates (1984) studied the effect of different solvent systems for extraction of neutral lipids, glycolipids, and phospholipids from millet seeds. Solvents used during the study were n-hexane/ether (80:20), methanol/water (85:15), $CHCl_3$/MeOH (2:1), ethanol/ether/water (2:2:1), ethanol/water (80:20), isopropanol and $CHCl_3$/MeOH (1:2), cold water-saturated butanol, and hot water-saturated butanol. The maximum recovery of lipids was observed in extraction phase hot water-saturated butanol

Table 4.1 Type and Concentration of Amino Acid in Different Types of Millet

Amino Acids	Pearl Millet	Foxtail Millet	Finger Millet	Proso Millet	Kodo Millet	References
Phenylalanine (g/Kg)	44–56	43–67	32–84	43–56	56–60	Bean et al. (2019), McDonough et al. (2000), Waniska and Rooney (2000)
Histidine (g/Kg)	18–26	17–40	3–40	18–29	15–22	
Isoleucine (g/Kg)	36–59	39–76	37–85	31–65	30–77	
Leucine (g/Kg)	80–251	114–136	64–162	106–154	67–127	
Lysine (g/Kg)	17–65	16–18	22–55	14–43	30–35	
Methionine (g/Kg)	15–29	16–31	13–43	13–26	15–18	
Threonine (g/Kg)	12–49	36–37	27–58	23–45	27–31	
Valine (g/Kg)	48–70	38–69	58–104	40–65	38–72	
Tryptophan (g/Kg)	11–28	4–19	10–17	6–17	5–10	
Aspartic acid (g/Kg)	49–103	64–77	57–100	37–63	61–64	
Glutamic acid (g/Kg)	123–254	176–199	203–378	149–223	122–339	
Alanine (g/Kg)	75–105	86–93	59–89	39–122	53–60	
Arginine (g/Kg)	32–81	27–95	34–82	27–91	36–50	
Cysteine (g/Kg)	7–28	5–15	7–29	5–28	7–10	
Glycine (g/Kg)	28–58	28–31	33–59	17–25	30–47	
Proline (g/Kg)	59–142	105–107	42–101	53–104	53–92	
Serine (g/Kg)	37–56	46–59	51–87	48–69	40–42	
Tyrosine (g/Kg)	17–48	22–32	20–56	18–40	34–42	

(7.19%). The amount of neutral lipids was maximum in $CHCl_3$/MeOH (2:1) (6.50%), whereas the maximum amounts of glycolipids and phospholipids were observed in methanol/water (85:15) (0.22%) and hot water-saturated butanol (0.84%), respectively.

4.2.4 Minerals and Vitamins

Minerals and vitamins are specific nutrients that are present in natural resources and are required for sustaining normal metabolic reactions in the body. The human body is not capable of synthesizing all the essential nutrients; therefore, it becomes necessary to provide these through specific food materials. It is important to add nutrient-rich food sources in the daily dietary menu to stay healthy and disease free. Each mineral performs specific functions for the welfare of body; for example, calcium is necessary for maintenance of teeth and bones and helps to regulate neural functioning and blood pressure. Iron is an essential mineral as it regulates enzymatic activation and helps in the synthesis of myoglobin, hemoglobin, hormones, and neurotransmitters. Inclusion of zinc in the diet helps to boost the immune system, aids in wound healing, and regulates cellular division. Millets are a good source of minerals and vitamins such as calcium, iron, phosphorus, zinc, thiamine, niacin, and riboflavin. However, their amount varies based on the types of millets. For instance, pearl millet had the maximum amount of phosphorus (3390 mg/Kg); foxtail millet is rich in zinc (606 mg/Kg) and thiamine (6 mg/Kg). Finger millet is a good source of calcium (3480 mg/Kg), whereas barnyard millet is rich in iron (174.7 mg/Kg) and riboflavin (42 mg/Kg) (Table 4.2). Proso millet, compared with other millets, is rich in niacin (45.4 mg/Kg) (Kumar et al., 2018; Chandel et al., 2014; Pontieri et al., 2014; Saleh et al., 2013; Panghal et al., 2006; Leder, 2004).

4.2.5 Fibers

Fibers are one of the important health-benefiting non-starchy carbohydrates required to sustain a healthy lifestyle. Two different terms are being used for the fibers: (1) dietary fibers and (2) functional fibers. The fibers that are naturally present in the food materials are known as dietary fibers, whereas those that are artificially added to food materials for obtaining health benefits are termed as functional fibers. On the solubility basis, fibers are basically of two different types: (1) soluble fibers

Table 4.2 Micronutrients (mg/kg) in Millets

Millets	Ca	Fe	P	Zn	Na	K	Mg	Mn	Cu	Thiamine	Niacin	Riboflavin	References
Pearl millet	350	103	3390	–	120	4420	1370	18	10.6	3	11.1	14.8	Kaur et al. (2019), Kumar et al. (2018), Chandel et al. (2014), Pontieri et al. (2014), Saleh et al. (2013), Panghal et al. (2006), Leder (2004), Serna-Saldivar and Rooney (1995), Barbeau and Hilu (1993)
Foxtail millet	310	35	3000	606	100	4000	1300	260	30	6	5.5	16.5	
Finger millet	3480	42.7	2500	366	110	5700	1370	55	40	4	8	6	
Barnyard millet	183.3	174.7	–	574.5	–	–	–	–	–	3.3	1	42	
Proso millet	100	22	2000	–	100	3200	1530	18.1	58	4.1	45.4	2.8	
Kodo millet	323.3	31.7	3000	327	100	1700	1660	29	58	1.5	0.9	20	

and (2) insoluble fibers. Generally, soluble fibers are present in significant amount in oats, fruits, and legumes, and they have the potential to reduce the level of low-density lipoproteins and also help in management of blood sugar level. Insoluble fibers are usually present in cereal grains, bran, and vegetables, and they play an important role in maintaining appetite and reducing the incidence of diabetes (type-2) and constipation. The fiber profile of millets indicates their presence in different millets as pearl millet (20.8%) and finger millet (18.6%) (Kamath & Belavady, 1980). Scientific studies demonstrated that pearl millet is a good source of both soluble and insoluble fibers whose amount varies from 2% to 4.5% (Ragaee et al., 2006; Ali et al., 2003). A significant amount of dietary fibers (37%–38%) were observed in kodo millet and little millet (Hegde & Chandra, 2005; Malleshi & Hadimani, 1993).

4.2.6 Bioactive Compounds

Bioactive compounds are important secondary metabolites that are well known for their health-benefiting antioxidant potential (Singh et al., 2019; Kaur et al., 2018). They are major contributors in pharmaceutical/ food industries for the synthesis of health-benefiting products. Bioactive compounds are documented for having antidiabetic, anticancerous, anti-microbial, and DNA damage-protection potentials (Dhull et al., 2020; Salar et al., 2017c; Siroha & Sandhu, 2017). While working with bioactive compounds, one of the crucial factors is their bioavailability. Once the bioactive components come out from the matrix of natural resources, they can be easily consumed for obtaining the maximum benefits. The governing factors during the extraction of bioactive constituents are extraction temperature, type of extraction phase, concentration, and time duration (Aguilera et al., 2019; Wu et al., 2017; Salar et al., 2016). Depending on the natural substrate being extracted during the process, the choice of extraction phase varies accordingly.

Millets are good source of health-benefiting specific bioactive compounds (Figure 4.1). Salar and Purewal (2017) demonstrated the variability in total phenolic compounds (TPC) and condensed tannin content (CTC) in 12 pearl millet cultivars. TPC and CTC in 12 cultivars were in the range of 3.8–7.32 mg GAE/g and 25.09–138.4 mg CE/100 g, respectively. Salar et al. (2017a) studied pearl millet cultivar PUSA-415 in terms of phenolic compounds and condensed tannin content. The extraction phase used during their study was ethanol (50%), and the amounts of TPC and

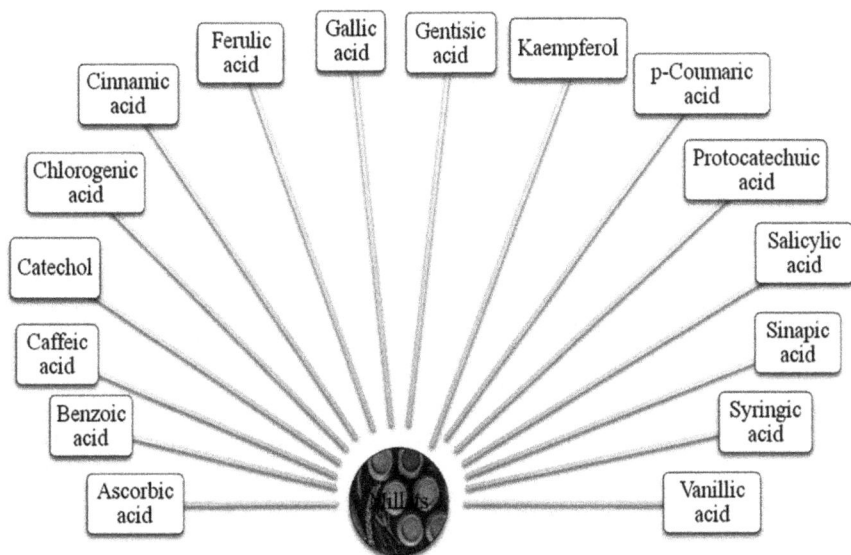

Figure 4.1 Bioactive compounds in millets.

CTC recovered were 6.4 mg GAE/g and 125.1 mg CE/100 g, respectively. Hithamani and Srinivasan (2014) studied finger millet and pearl millet in terms of phenolic compounds using acidified methanol as extraction phase. They observed that TPC in both millets ranged from 5.95 to 11.9 mg/g. Chandrasekara et al. (2012) studied different millets (foxtail, proso, finger, kodo, little, and pearl millets) in terms of phenolic compounds and identified TPC amounts of 7.1–32.4 µmol FAE/g. Pradeep and Guha (2011) reported 4.29–5.21 mg GAE/g TPC in little millet. During their study, acidified methanol was used as the extraction phase to recover phenolics from little millet. Chandrasekara and Shahidi (2011b) used acetone (70%) to extract phenolics from foxtail, proso, finger, kodo, little, and pearl millets. TPC was observed in range of 146.3–1156.6 µmol FAE/g. Sharma et al. (2016) studied foxtail millet, barnyard millet, and kodo millet in terms of phenolic compounds and observed TPC in the range of 23.4–62.2 mg/100 g. Sharma et al. (2017) demonstrated that extraction of kodo millet with water, methanol, and acetone resulted in recovery of phenolic compounds, which ranged from 79 to 359.2 mg/100 g. Obadina et al. (2016) reported the amount of TPC in pearl millet (native and processed) as 90.6–169.8 mg/100 g.

Pradeep and Sreerama (2017) studied foxtail and little millets for the detection of phenolic compounds in them. TPC was observed in millet extracts, ranging from 2.8 to 34.6 mg FAE/g. Marmouzi et al. (2018) detected phenolic compounds in pearl millet and found them in the range of 4.1–22.7 mg GAE/g. Pradeep and Sreerama (2015) reported 73.6–246.8 mg FAE/100 g TPC in barnyard, foxtail, and proso millets. The extraction conditions used during their study were a combination of HCl-methanol (1%) for a duration of 30 min. Devisetti et al. (2014) demonstrated that foxtail and proso millets could be good sources of phenolics compounds as indicated by TPC in the range of 0.24–1.83 mg/g. HPLC analysis of millets indicates the presence of apigenin, ascorbic acid, caffeic acid, catechin, catechin-O-dihexoside, cinnamic acid, epicatechin, epigallocatechin, ferulic acid, gallic acid, kempherol, p-coumaric acid, p-hydroxy benzoic acid, procyanidin dimer, protocatechuic acid;, protocatechuic aldehyde, quercetin, rutin, sinapic acid, syringic acid, and vanillic acid (Xiang et al., 2018; Shahidi & Chandrasekara, 2013; Banerjee et al., 2012; Chandrasekara & Shahidi, 2011; Shobana et al., 2009; Viswanath et al., 2009; Chethan et al., 2008; Rao & Muralikrishna, 2002). The presence of specific bioactive compounds in millets makes them an attractive substrate at the industrial scale for the preparation of various health-benefiting functional food products.

4.3 HEALTH BENEFITS OF MILLETS

Being a good source of starch, protein, fiber, minerals, vitamins, and bioactive compounds, millets could be a potential substrate for the preparation of health-benefiting food products. Bioactive compounds extracted from millets are industrially important and could be useful in the treatment of various diseases (Figure 4.2).

4.3.1 Diabetes and Celiac Disease

Worldwide, diabetes is the most prevalent disease affecting the daily routine life of the majority of people. Diabetic people have to more conscious about what to eat and when to eat. A balanced diet is necessary to keep them healthy. Besides medication, the selection of suitable food materials and changes in lifestyle may help diabetic persons survive. The risk of diabetes symptoms could be minimized using a suitable diet, correlated with the severity/risk (Greenwood et al., 2013). From the health point of view,

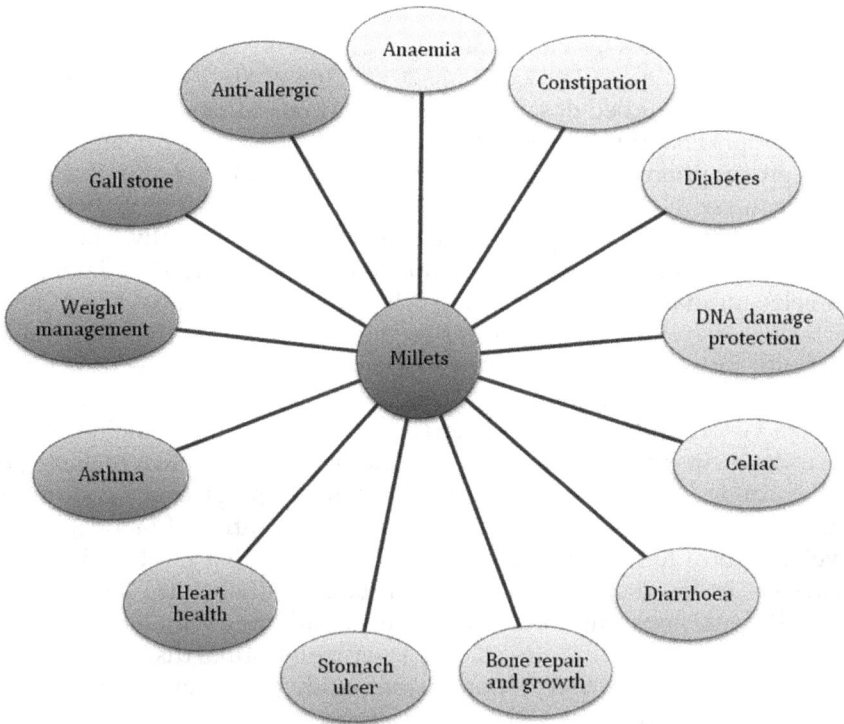

Figure 4.2 Medical uses of millets.

dietary fibers play an important role in the body as they help in boosting the functionality during gastrointestinal reactions and decrease the rate of glucose increase in serum. Studies on millets demonstrated that their regular consumption results in significant management of cholesterol level (Sharma & Niranjan, 2017; Bangoura et al., 2013; Anderson et al., 2009). Bangoura et al. (2013) reported that insoluble fibers present in foxtail millet may delay glucose diffusion as well as promote their absorption in the intestinal tract. Consumption of millets and millet-based food products inhibits the activity of α-amylase, which ultimately affects the carbohydrate's digestibility and slows down the release of glucose. The function performed by millets and their products makes them an ideal food source for improving sensitivity of insulin in diabetes mellitus (Ju-Sung et al., 2011). Along with the management of insulin and glucose levels, millets are also helpful in managing elevated cholesterol level and reduce the chances of cardiovascular disorders (Sharma & Niranjan, 2017; Sireesha et al., 2011; Lee et al., 2010; Choi et al., 2005). Kumari and Sumathi (2002) studied dosa prepared from finger millet and observed its effect on plasma glucose level. The outcome of their study demonstrated that consumption of dosa prepared from finger millet results in significant decrease in plasma glucose level with less peak rise. In diabetes, the capacity of the body to repair wounds decreases dramatically. The abnormality in physiological response is due to free radical-assisted damage to the cells, ultimately resulting in necrosis and deeper wounds (King, 2001). Inclusion of antioxidant-rich foods in the diet increases the potential of the body for wound healing. Rajasekaran et al. (2004) studied the effect of finger millet consumption on diabetic rats for the production of nerve growth factors and wound healing. They observed that diet including finger millet helps in enhancement of wound healing capability. Inclusion of millets in daily diet is suggested to overcome hyperglycemic conditions (insufficient secretion of insulin). Postprandial control of blood glucose level is critical for diabetic patients; however, addition of fiber-rich food materials in the diet could maintain a balance (Lebovitz, 2001; Baron, 1998). Regular consumption of millets, especially finger millets, is associated with low risk of diabetes as well as gastrointestinal disorders (Tovey, 1994). Sireesha et al. (2011) reported the anti-hyperglycemic potential of the foxtail millet aqueous extracts in streptozotocin-induced diabetic rats. They observed that a specific amount of extract dose (300 mg/Kg) resulted in a significant decrease (70%) in blood glucose level of diabetic rats. Choi et al. (2005) demonstrated that foxtail millet has the potential to manage cholesterol and insulin levels. Lee et al. (2010) observed that consumption of millets results in regulation of protein concentration (C-reactive) and lipid level, with a decrease in triglyceride level.

Millet grains are well documented for their health-benefiting phenolic compounds, tannins, and antioxidant potential (Sandhu et al., 2020; Shahidi & Chandrasekara, 2013; Chandrasekara & Shahidi, 2010). Consumption of millets is reported to be involved in decreasing the chances of cancer initiation, especially esophageal cancer (Van Rensburg, 1981). One of the main reasons behind the increasing popularity of millets is their gluten-free nature. Worldwide, millets are gaining interest because the products based on them are tasty and health benefiting. Celiac disease is one of the problems faced by people throughout the world (Catassi & Fasano, 2008). The disease is an autoimmune disorder in which the gluten ingestion poses damages to the small intestine. The gluten-free nature of millet-based products could be a boon for the patient suffering from celiac disease (Taylor & Emmambux, 2008; Taylor et al., 2006).

4.3.2 Antimicrobial Potential

Antimicrobial agents are those that have potential either to control or kill the microbial strains, like fungi or bacteria (Rani et al., 2014). The antimicrobial potential of any natural resource could be evaluated on the basis on zone of inhibition against any microbial strains. During the antimicrobial assay, a suitable amount of extract (e.g., 100 or 150 µl) is added into the slot created using a cork borer in the nutrient media. The nutrient media provides favorable conditions for the growth of microorganisms. If the extract has any metabolites that can inhibit the growth of specific microorganisms, the results could be easily observed with the naked eye in the form of zone of inhibition (the area in which microorganisms fails to grow). Fractions of millets and extracts prepared from them are well documented for their antimicrobial potential. Viswanath et al. (2009) demonstrated that extracts prepared from the seed coat of finger millet have significant antimicrobial potential against *B. cereus* and *A. flavus*. Their findings support the concept that addition of finger millet in the diet could be health benefiting. Bisht et al. (2016) studied proso millet and finger millet in terms of antimicrobial potential. They observed that, compared with finger millet (22.6 mm), proso millet has the least antimicrobial potential (15.6 mm) against gram-positive bacteria. Amadou et al. (2013) studied the antimicrobial potential of finger millet extracts against *Escherichia coli*, *Salmonella typhi*, and *Staphylococcus aureus*. The zones of inhibition they observed during the antimicrobial assay were 18.4 mm against *E. coli*, 17.3 mm for *S. aureus*, and 17 mm for *S. typhi*. Ogbeba (2019) reported antimicrobial activity of finger millet extracts against *P. aeruginosa*, *E. coli*, and *S. aureus*. The outcome of the study indicates that different concentrations of finger millet extracts at concentrations of 100 mg/ml showed zones of inhibition against *P. aeruginosa* (8 mm), *S. aureus* (5 mm), and *E. coli* (4 mm).

4.3.3 Bone Health and DNA Damage Protection

More than 20 minerals are required by the body to perform various metabolic reactions to sustain healthy life. Out of those 20 minerals, calcium is one of the important minerals required for maintaining healthy bones.

Millets, especially finger millet, is a good source of calcium whose consumption helps the person to combat osteopenia and osteoporosis. Fortification of millet flour in bakery products could be an alternate method to provide calcium in adequate amount. Nordin (1976) demonstrated that in adults the proportion of calcium is approximately 1.9% of total body weight. Calcium is solely required for normal muscular contraction, mediation of the transmission of neural signaling, and bone formation. Sharma et al. (2017) recommend the use of millets in daily routine diet as it can fulfill the required amount of specific minerals and calcium.

DNA is one of the widely studied molecules to understand recombination techniques, diseases, modifications, sequencing, and agroindustrial growth for human welfare (Liu, 2017; Sawitzke et al., 2017). Oxidative stress is one of the factors that may result in damage at the DNA level. During cellular injury and DNA damage, the body starts to release cytokines, which results in signaling (Mantovani et al., 2008). Damage to DNA may occur by two different ways: one is exogenous damage to DNA, and the other is endogenous DNA damage (Kryston et al., 2011). Extracts prepared using different extraction phases are widely being studied for the presence of DNA damage protection activity (Salar & Purewal, 2016; Xiao et al., 2015). During in vitro assays, a well-known Fenton reagent is allowed to react with DNA and cause damage in it. Further, extracts were evaluated for the capability to slow down/stop the damaging effect of the Fenton reagent. Presence of DNA damage protection activity in millets extract makes them a pharmacologically important crop as it may help in developing specific vaccines and medicinally important compounds that may combat DNA damaging reactions within the body.

4.3.4 Other Health Benefits

The mineral profile of millets demonstrated that grains are good sources of minerals, especially potassium and magnesium, which play important roles in maintaining blood pressure and reducing the risk of strokes/heart attacks. Antioxidant properties in millets help to combat cancer, diabetes, and chronic disorders. Further, fibers, minerals, and other important nutrients present in millets help remove toxic compounds from body. Fiber-rich food materials help in bowel functioning, which maintain routine digestion and absorption process (Schneeman & Tietyen, 1994). The fibrous nature of millet grains reduces the formation of gall stones and is helpful in maintaining body weight.

4.4 CONCLUSION

Being gluten free, millets could be recommended for a majority of people suffering from celiac disease. To meet the increasing demand of health-benefiting food products, millets could serve as suitable substrates. Millet grains are easily available on the market at affordable prices; so, they could be industrially used for the preparation of various processed formulations. The presence of specific compounds with antioxidant properties makes millets an important cereal grain that is pharmacologically important. Regular consumption of millets can help eradicate the occurrence of disease conditions. Every age group could use millets and millet-based bakery products for sustaining healthy life.

REFERENCES

Adebowale, A. R. A., Emmambux, M. N., Beukes, M. and Taylor, J. R. N. 2011. Fractionation and characterization of teff proteins. *Journal of Cereal Science* 54: 380–386.

Aguilera, Y., Rebollo-Hernanz, M., Canas, S., Taladrid, D. and Martin-Cabrejas, M. 2019. Response surface methodology to optimise the heat-assisted aqueous extraction of phenolic compounds from coffee parchment and their comprehensive analysis. *Food & Function* 10: 4739–4750.

Ali, M. A. M., El-Tinay, A. H. and Abdalla, A. H. 2003. Effect of fermentation on the in vitro protein digestibility of pearl millet. *Food Chemistry* 80(1): 51–54.

Amadou, I., Le, G. W. and Shi, Y. H. 2013. Evaluation of antimicrobial, antioxidant activities, and nutritional values of fermented foxtail millet extracts by *Lactobacillus paracasei* Fn032. *International Journal of Food Properties* 16: 1179–1190.

Anderson, J. W., Baird, P., Davis, R. H., Ferreri, S., Knudtson, M., Koraym, A., Waters, V. and Williams, C. L. 2009. Health benefits of dietary fiber. *Nutrition Reviews* 67: 188–205.

Annor, G. A., Marcone, M., Bertoft, E. and Seetharaman, K. 2014. Physical and molecular characterization of millet starches. *Cereal Chemistry* 91: 286–292.

Balasubramanian, S., Sharma, R., Kaur, J. and Bhardwaj, N. 2011. Isolation, modification and characterization of finger millet (*Eleucine coracana*) starch. *Journal of Food Science & Engineering* 1: 339–347.

Banerjee, S., Sanjay, K., Chethan, S. and Malleshi, N. G. 2012. Finger millet (*Eleusine coracana*) polyphenols: in vestigation of their antioxidant capacity and antimicrobial activity. *African Journal of Food Science* 6: 362–374.

Bangoura, M. L., Nsor-Atindana, J., Zhu, K., Tolno, M. B., Zhou, H. and Wei, P. 2013. Potential hypoglycaemic effects of insoluble fibres isolated from foxtail millets (*Setaria italica* (L.) P. Beauvois). *International Journal of Food Science and Technology* 48: 496–502.

Barbeau, W. E. and Hilu, K. W. 1993. Protein, calcium, iron, and amino acid content of selected wild and domesticated cultivars of finger millet. *Plant Foods for Human Nutrition* 43: 97–104.

Baron, A. D. 1998. Postprandial hyperglycemia and alpha-glucosidase inhibitors. *Diabetes Research and Clinical Practice* 40: S51–S55.

Bean, S. R., Zhu, L., Smith, B. M., Wilson, J. D., Loerger, B. P. and Tilley, M. 2019. Starch and protein chemistry and functional properties. In: Taylor, J. N. and Duodu, K. G., editor. *Sorghum and millets (second edition) chemistry, technology and nutritional attributes*. Duxford, United Kingdom: Woodhead Publishers, pp. 131–170.

Bisht, A., Thapliyal, M. and Singh, A. 2016. Screening and isolation of antibacterial proteins/peptides from seeds of millets. *International Journal of Current Pharmaceutical Research* 8: 96–99.

Boyles, R. E., Pfeiffer, B. K., Cooper, E. A., Rauh, B. L., Zielinski, K. J., Myers, M. T., Brenton, Z., Rooney, W. L. and Kresovich, S. K. 2017. Genetic dissection of sorghum grain quality traits using diverse and segregating populations. *Theoretical & Applied Genetics* 130: 697–716.

Brasil, T. A., Capitani, C. D., Takeuchi, K. P. and Ferreira, T. A. P. de C. 2015. Physical, chemical and sensory properties of gluten-free kibbeh formulated with millet flour (*Pennisetum glaucum* (L.) R. Br.). *Food Science and Technology (Campinas)* 35: 361–367.

Catassi, C. and Fasano, A. 2008. Celiac disease. In: Gallagher, E., editor. *Gluten-free cereal products and beverages*. Burlington, MA: Elsevier. pp. 1–27.

Chandel, G., Kumar, M., Dubey, M. and Kumar, M. 2014. Nutritional properties of minor millets: neglected cereals with potentials to combat malnutrition. *Current Science* 107: 1109–1111.

Chandrasekara, A., Naczk, M. and Shahidi, F. 2012. Effect of processing on the antioxidant activity of millet grains. *Food Chemistry* 133: 1–9.

Chandrasekara, A. and Shahidi, F. 2010. Content of insoluble bound phenolics in millets and their contribution to antioxidant capacity. *Journal of Agricultural & Food Chemistry* 58: 6706–6714.

Chandrasekara, A. and Shahidi, F. 2011a. Determination of antioxidant activity in free and hydrolyzed fractions of millet grains and characterization of their phenolic profiles by HPLC-DAD-ESI-MSn. *Journal of Functional Foods* 3: 144–158.

Chandrasekara, A. and Shahidi, F. 2011b. Antiproliferative potential and DNA scission inhibitory activity of phenolics from whole millet grains. *Journal of Functional Foods* 3: 159–170.

Chethan, S., Dharmesh, S. M. and Malleshi, N. G. 2008. Inhibition of aldose reductase from cataracted eye lenses by finger millet (*Eleusine coracana*) polyphenols. *Bioorganic & Medicinal Chemistry* 16: 10085–10090.

Choi, Y. Y., Osada, K., Ito, Y., Nagasawa, T., Choi, M. R. and Nishizawa, N. 2005. Effects of dietary protein of Korean foxtail millet on plasma adiponectin, HDL-cholesterol, and insulin levels in genetically type 2 diabetic mice. *Bioscience, Biotechnology, & Biochemistry* 69: 31–37.

Cremer, J. E., Bean, S. R., Tilley, M., Ioerger, B. P., Ohm, J. B., Kaufmann, R. C., Wilson, J. D., Innes, D. D., Gilding, E. K. and Godwin, I. D. 2014. Grain sorghum proteomics: an integrated approach towards characterization of seed storage proteins in kafirin allelic variants. *Journal of Agricultural & Food Chemistry* 62: 9819–9831.

Devi, P. B., Vijayabharathi, R., Sathyabama, S., Malleshi, N. G. and Priyadarisini, V. B. 2011. Health benefits of finger millet (*Eleusine coracana* L.) polyphenols and dietary fiber: a review. *Journal of Food Science & Technology* 51: 1021–1040.

Devisetti, R., Yadahally, S. N. and Bhattacharya, S. 2014. Nutrients and antinutrients in foxtail and proso millet milled fractions: evaluation of their flour functionality. *LWT-Food Science &Technology* 59: 889–895.

Dhull, S. B., Punia, S., Kidwai, M. K., Kaur, M., Chawla, P., Purewal, S. S., Sangwan, M. and Palthania, S. 2020. Solid-state fermentation of lentil (*Lens culinaris* L.) with *Aspergillus awamori*: effect on phenolic compounds, mineral content, and their bioavailability. *Legume Science*: 1–12.

Geervani, P. and Eggum, B. O. 1989. Nutrient composition and protein quality of minor millets. *Plant Foods for Human Nutrition* 39: 201–208.

Greenwood, D. C., Threapleton, D. E., Evans, C. E. L., Cleghorn, C. L., Nykjaer, C., Woodhead, C. and Burley, V. J. 2013. Glycemic index, glycemic load, carbohydrates, and Type 2 diabetes: systematic review and dose response meta-analysis of prospective studies. *Diabetes Care* 36: 4166–4171.

Hegde, P. S. and Chandra, T. S. 2005. ESR spectroscopic study reveals higher free radical quenching potential in kodo millet (*Paspalum scrobiculatum*) compared to other millets. *Food Chemistry* 92: 177–182.

Himashu, Chauhan, M., Sonawane, S. K. and Arya, S. S. 2018. Nutritional and nutraceutical properties of millets: a review. *Clinical Journal of Nutrition & Dietetics* 1: 1–10.

Hithamani, G. and Srinivasan, K. 2014. Effect of domestic processing on the polyphenol content and bioaccessibility in finger millet (*Eleusine coracana*) and pearl millet (*Pennisetum glaucum*). *Food Chemistry* 164: 55–62.

Hooks, T., Pedersen, J. F., Marx, D. B. and Vogel, K. P. 2006. Variation in the U.S. photoperiod insensitive sorghum collection for chemical and nutritional traits. *Crop Science* 46: 751–757.

Indira, R. and Naik, M. S. 1971. Nutrient composition and protein quality of some minor millets. *Indian Journal of Agriculture Science* 41: 795–797.

Izumi, Y., Okaichi, S., Awala, S. K., Kawato, Y., Watanabe, Y., Yamane, K. and Iijima, M. 2018. Water supply from pearl millet by hydraulic lift can mitigate drought stress and improve productivity of rice by the close mixed planting. *Plant Production Science* 21: 8–15.

Ju-Sung, K., Tae Kyung, H. and Myong, Jo K. 2011. The inhibitory effects of ethanol extracts from sorghum, foxtail millet and proso millet on alpha-glucosidase and alpha-amylase activities. *Food Chemistry* 124: 1647–1651.

Kamath, M. V. and Belavady, B. 1980. Unavailable carbohydrates of commonly consumed Indian foods. *Journal of the Science of Food & Agriculture* 31: 192–202.

Kaur, R., Kaur, M. and Purewal, S. S. 2018. Effect of incorporation of flaxseed to wheat rusks: antioxidant, nutritional, sensory characteristics, and in vitro DNA damage protection activity. *Journal of Food Processing & Preservation* 42: 13585.

Kaur, P., Purewal, S. S., Sandhu, K. S., Kaur, M. and Salar, R. K. 2019. Millets: a cereal grain with potent antioxidants and health benefits. *Journal of Food Measurement & Characterization* 13: 793–806.

Kim, S. K., Sohn, E. Y. and Lee, I. J. 2009. Starch properties of native foxtail millet, *Setaria italica* Beauv. *Journal of Crop Science & Biotechnology* 12: 59–62.

King, L. 2001. Impaired wound healing in patients with diabetes. *Nursing Standard* 15: 39–45.

Kryston, T. B., Georgiev, A. B., Pissis, P. and Georgakilas, A. G. 2011. Role of oxidative stress and DNA damage in human carcinogenesis. *Mutation Research* 711: 193–201.

Kumar, K. K. and Paramswaran, K. P. 1998. Characterisation of storage protein from selected varieties of foxtail millet (*Setaria italic* (L) Beauv). *Journal of the Science of Food and Agriculture* 77: 535–542.

Kumar, A., Tomer, V., Kaur, A., Kumar, V. and Gupta, K. 2018. Millets: a solution to agrarian and nutritional challenges. *Agriculture & Food Security* 7: 1–15.

Kumari, L. P. and Sumathi, S. 2002. Effect of consumption of finger millet on hyperglycemia in non-insulin dependent diabetes mellitus (NIDDM) subjects. *Plant Foods for Human Nutrition* 57: 205–213.

Kumari, K. S. and Thayumanavan, B. 1998. Characterization of starches of proso, foxtail, barnyard, kodo, and little millets. *Plant Foods for Human Nutrition* 53: 47–56.

Lebovitz, H. E. 2001. Effect of the postprandial state on non traditional risk factors. *American Journal of Cardiology* 88: 204–205.

Leder, I. 2004. Sorghum and millets. cultivated plants, primarily as food sources. In: Gyargy, F., editor. *Encyclopedia of life support systems.* Oxford: UNESCO, Eolss Publishers.

Lee, S. H., Chung, I. M., Cha, Y. S. and Park, Y. 2010. Millet consumption decreased serum concentration of triglyceride and C-reactive protein but not oxidative status in hyperlipidemic rats. *Nutrition Research* 30: 290–296.

Liu, S. 2017. *Chapter-13: Evolution and genetic engineering. Bioprocess engineering.* 2nd ed. pp. 783–828.

Liu, C., Liu, P., Yan, S., Qing, Z. and Shen, Q. 2010. Relationship of physicochemical, pasting properties of millet starches and the texture properties of cooked millet. *Journal of Texture Studies* 42: 247–253.

Malleshi, N. G. and Hadimani, N. A. 1993. Nutritional and technological characteristics of small millets and preparation of value-added products from them. In: Riley, K. W., Gupta, S. C., Seetharam, A. and Mushonga, J. N., editors. *Advances in small millets.* New Delhi: Oxford and IBH Publishing Co Pvt. Ltd. pp. 271–287.

Mantovani, A., Allavena, P., Sica, A. and Balkwill, F. 2008. Cancer-related inflammation. *Nature* 454: 436–444.

Marmouzi, I., Ali, K., Harhar, H., Gharby, S., Sayah, K., El Madani, N., Cherrah, Y. and Faouzi, M. E. A. 2018. Functional composition, antibacterial and antioxidative properties of oil and phenolics from Moroccan Pennisetum glaucum seeds. *Journal of the Saudi Society of Agricultural Sciences* 17: 229–234.

McDonough, C. M., Rooney, L. W. and Serna-Saldivar, S. O. 2000. The millets. In: Kulp, K. K., editor, *Handbook of cereal science and technology*. 2nd ed. New York, NY: Marcel Dekker. pp. 177–202.

Murray, S. C., Sharma, A., Rooney, W. L., Klein, P. C., Mullet, J. E., Mitchell, S. E. and Kresovich, S. 2008. Genetic improvement of sorghum as a biofuel feedstock: I. QTL for stem sorghum and grain nonstructural carbohydrates. *Crop Science* 48: 2165–2179.

Nambiar, V. S., Dhaduk, J. J., Sareen, N., Shahu, T. and Desai, R. 2011. Potential functional implications of pearl millet (*Pennisetum glaucum*) in health and disease. *Journal of Applied Pharmaceutical Science* 1: 62–67.

Nordin, B. E. C. 1976. Nutritional considerations. In Nordin, B. E. C., editor. *Calcium, phosphate and magnesium metabolism*. Edinburgh: Churchill Livingstone. pp. 1–35.

Obadina, A., Ishola, I. O., Adekoya, I. O., Soares, A. G., de Carvalho, C. W. P. and Barboza, H. T. 2016. Nutritional and physico-chemical properties of flour from native and roasted whole grain pearl millet (*Pennisetum glaucum* [L.]R. Br.). *Journal of Cereal Science* 70: 247–252.

Ogbeba, J. 2019. Phytochemical and antibacterial property of finger millet (*Eleusine coracana*) on some selected clinical bacteria. *Access Microbiology* 1: 135.

Osagie, A. U. and Kates, M. 1984. Lipid composition of millet (*Pennisetum americanum*) seeds. *Lipids* 19: 958–965.

Panghal, A., Khatkar, B. S. and Singh, U. 2006. Cereal proteins and their role in food industry. *Indian Food Industry* 25: 58–62.

Pontieri, P., Troisi, J., Fiore, R. D., Bean, S. R., Roemer, E., Bofa, A., Giudice, A. D., Pizzolante, G., Alifano, P. and Giudice, L. D. 2014. Mineral contents in grains of seven food grade sorghum hybrids grown in mediterranean environment. *Australian Journal of Crop Science* 8: 1550–1559.

Pradeep, S. R. and Guha, M. 2011. Effect of processing methods on the nutraceutical and antioxidant properties of little millet (*Panicum sumatrense*) extracts. *Food Chemistry* 126: 1643–1647.

Pradeep, P. M. and Sreerama, Y. N. 2015. Impact of processing on the phenolic profiles of small millets: evaluation of their antioxidant and enzyme inhibitory properties associated with hyperglycemia. *Food Chemistry* 169: 455–463.

Pradeep, P. M. and Sreerama, Y. N. 2017. Soluble and bound phenolics of two different millet genera and their milled fractions: comparative evaluation of antioxidant properties and inhibitory effects on starch hydrolysing enzyme activities. *Journal of Functional Foods* 35: 682–693.

Punia, S., Sandhu, K. S., Dhull, S. G., Siroha, A. K., Purewal, S. S., Kaur, M. and Kidwai, M. K. 2020. Oat starch: physico-chemical, morphological, rheological characteristics and its application–a review. *International Journal of Biological Macromolecules* 154: 493–498.

Punia, S., Siroha, A. K., Sandhu, K. S. and Kaur, M. 2019. Rheological and pasting behavior of OSA modified mungbean starches and its utilization in cake formulation as fat replacer. *International Journal of Biological Macromolecules* 128: 230–236.

Purewal, S., Salar, R., Bhatti, M., Sandhu, K. S., Singh, S. K. and Kaur, P. 2020. Solid-state fermentation of pearl millet with *Aspergillus oryzae* and *Rhizopus azygosporus*: effects on bioactive profile and DNA damage protection activity. *Journal of Food Measurement and Characterization* 14: 150–162.

Purewal, S. S., Sandhu, K. S., Salar, R. K. and Kaur, P. 2019. Fermented pearl millet: a product with enhanced bioactive compounds and DNA damage protection activity. *Journal of Food Measurement & Characterization* 13: 1479–1488.

Ragaee, S., Abdel-Aal, E. M. and Noaman, M. 2006. Antioxidant activity and nutrient composition of selected cereals for food use. *Food Chemistry* 98: 32–38.

Rajasekaran, N. S., Nithya, M., Rose, C. and Chandra, T. S. 2004. The effect of finger millet feeding on the early responses during the process of wound healing in diabetic rats. *Biochimica et Biophysica Acta* 1689: 190–201.

Rani, R., Kumar, H., Salar, R. K. and Purewal, S. S. 2014. Antibacterial activity of copper oxide nanoparticles against gram-negative bacterial strain synthesized by reverse micelle technique. *International Journal of Pharmaceutical Research & Development* 6: 072–078.

Rao, M. V. S. S. T. S. and Muralikrishna, G. 2002. Evaluation of the antioxidant properties of free and bound phenolic acids from native and malted finger millet (Ragi, Elucine coracana Indaf-15). *Journal of Agricultural and Food Chemistry* 50: 889–892.

Rhodes, D. H., Hoffman, Jr., Rooney, W. L., Herald, T. J., Bean, S., Boyles, R., Brenton, Z. W. and Kresovitch, S. 2017. Genetic architecture of kernel composition in global sorghum germplasm. *BMC Genomics* 18: 15.

Rockström, J. and de Rouw, A. 1997. Water, nutrients and slope position in on-farm pearl millet cultivation in the Sahel. *Plant and Soil* 195: 311–327.

Salar, R. K. and Purewal, S. S. 2016. Improvement of DNA damage protection and antioxidant activity of biotransformed pearl millet (*Pennisetum glaucum*) cultivar PUSA-415 using *Aspergillus oryzae* MTCC 3107. *Biocatalysis & Agricultural Biotechnology* 8: 221–227.

Salar, R. K. and Purewal, S. S. 2017. Phenolic content, antioxidant potential and DNA damage protection of pearl millet (*Pennisetum glaucum*) cultivars of North Indian region. *Journal of Food Measurement & Characterization* 11: 126–133.

Salar, R. K., Purewal, S. S. and Bhatti, M. S. 2016. Optimization of extraction conditions and enhancement of phenolic content and antioxidant activity of pearl millet fermented with *Aspergillus awamori* MTCC-548. *Resource Efficient Technologies* 2: 148–157.

Salar, R. K., Purewal, S. S. and Sandhu, K. S. 2017a. Fermented pearl millet (*Pennisetum glaucum*) with in vitro DNA damage protection activity, bioactive compounds and antioxidant potential. *Food Research International* 100: 204–210.

Salar, R. K., Purewal, S. S. and Sandhu, K. S. 2017b. Relationships between DNA damage protection activity, total phenolic content, condensed tannin content and antioxidant potential among Indian barley cultivars. *Biocatalysis & Agricultural Biotechnology* 11: 201–206.

Salar, R. K., Purewal, S. S. and Sandhu, K. S. 2017c. Bioactive profile, free-radical scavenging potential, DNA damage protection activity, and mycochemicals in *Aspergillus awamori* (MTCC 548) extracts: a novel report on filamentous fungi. *3 Biotech* 7: 164.

Saleh, A. S. M., Zhang, Q., Chen, J. and Shen, Q. 2013. Millet grains: nutritional quality, processing, and potential health benefits. *Comprehensive Reviews in Food Science & Food Safety* 12: 281–295.

Sandhu, K. S., Kaur, P., Siroha, A. K. and Purewal, S. S. 2020. Phytochemicals and antioxidant properties in pearl millet. In: Punia, S., Siroha, A. K., Sandhu, K. S., Gahlawat, S. K. and Kaur, M., editors. *Pearl millet: properties, functionality and its applications.* 1st ed., Boca Raton, London, New York: CRC Press. pp. 33–50.

Sandhu, K. S. and Siroha, A. K. 2017. Relationships between physicochemical, thermal, rheological and in vitro digestibility properties of starches from pearl millet cultivars. *LWT - Food Science &Technology* 83: 213–224.

Sawitzke, J. A., Thomason, L. C., Costantino, N. and Court, D. L. 2017. Recombineering: a modern approach to genetic engineering. *Reference Module in Life Science* 2013: 109–112.

Schneeman, B. O., Tietyen, J. 1994. Dietary fiber. In: Shills, M. E., Olson, J. A. and Shike, M., editors. *Modern nutrition in health and disease.* Philadelphia, PA: Lea and Febiger. pp: 89–100.

Serna-Saldivar, S. and Rooney, L. W. 1995. Structure and chemistry of sorghum and millets. In: Dendy, D. A. V., editor, *Sorghum and millets: chemistry and technology.* Saint Paul, MN: AACC Press. pp. 69–124.

Shahidi, F. and Chandrasekara, A. 2013. Millet grain phenolics and their role in disease risk reduction and health promotion: a review. *Journal of Functional Foods* 5: 570–581.

Sharma, D., Jamra, G., Singh, U. M., Sood, S. and Kumar, A. 2017. Calcium biofortification: three pronged molecular approaches for dissecting complex trait of calcium nutrition in finger millet (*Eleusine coracana*) for devising strategies of enrichment of food crops. *Frontiers in Plant Sciences* 7: 2028.

Sharma, N. and Niranjan, K. 2017. Foxtail millet: properties, processing, health benefits, and uses. *Food Reviews International* 34: 329–363.

Sharma, S., Saxena, D. C. and Riar, C. S. 2016. Nutritional, sensory and in-vitro antioxidant characteristics of gluten free cookies prepared from flour blends of minor millets. *Journal of Cereal Science* 72: 153–161.

Shobana, S., Sreerama, Y. N. and Malleshi, N. G. 2009. Composition and enzyme inhibitory properties of finger millet (*Eleusine coracana* L.) seed coat phenolics: mode of inhibition of aglucosidase and pancreatic amylase. *Food Chemistry* 115: 1268–1273.

Singh, S., Kaur, M., Sogi, D. S. and Purewal, S. S. 2019. A comparative study of phytochemicals, antioxidant potential and in-vitro DNA damage protection activity of different oat (*Avena sativa*) cultivars from India. *Journal of Food Measurement and Characterization* 13: 347–356.

Singh, P. and Raghuvanshi, S. 2012. Finger millet for food and nutritional security. *African Journal of Food Science* 6: 77–84.

Singh, P., Singh, U., Eggum, B. O., Kumar, K. A. and Andrews, D. J. 1987. Nutritional evaluation of high protein genotypes of pearl millet (*Pennisetum americanum* (L.) Leeke). *Journal of the Science of Food & Agriculture* 38: 41–48.

Sireesha, Y., Kasetti, R. B., Nabi, S. A., Swapna, S. and Apparao, C. 2011. Antihyperglycemic and hypolipidemic activities of *Setaria italica* seeds in STZ diabetic rats. *Pathophysiology* 18: 159–164.

Siroha, A. K., Punia, S., Purewal, S. S., Sharma, L. and Singh, A. 2020. Impact of different modifications on starch properties. In: Punia, S., Siroha, A. K., Sandhu, K. S., Gahlawat, S. K. and Kaur, M., editors. *Pearl millet: properties, functionality and its applications.* 1st ed.: CRC Press, CRC Press, Boca Raton, London, New York. pp. 91–114.

Siroha, A. K. and Sandhu, K. S. 2017. Effect of heat processing on the antioxidant properties of pearl millet (*Pennisetum glaucum* L.) cultivars. *Journal of Food Measurement and Characterization* 11: 872–878.

Siroha, A. K., Sandhu, K. S. and Kaur, M. 2016. Physicochemical, functional and antioxidant properties of flour from pearl millet varieties grown in India. *Journal of Food Measurement & Characterization* 10: 311–318.

Siroha, A. K., Sandhu, K. S., Kaur, M. and Kaur, V. 2019. Physicochemical, rheological, morphological and in vitro digestibility properties of pearl millet starch modified at varying levels of acetylation. *International Journal of Biological Macromolecules* 131: 1077–1083.

Subramanian, V., Seetharama, N., Jambunathan, R. and Rao, P. V. 1990. Evaluation of protein quality of sorghum [*Sorghum bicolor* (L.) Moench]. *Journal of Agricultural & Food Chemistry* 38: 1344–1347.

Sukumaran, S., Xiang, W., Bean, S. R., Pedersen, J. F., Tuinstra, M. R., Tesso, T. T., Hamblin, M. T. and Yu, J. 2012. Genetic structure of a diverse sorghum collection and association mapping for grain quality. *Plant Genome* 5: 126–135.

Taylor, J. R. N. and Emmambux, M. N. 2008. Gluten-free foods and beverages from millets. In: Gallagher, E., editor. *Gluten-free cereal products and beverages.* Burlington, MA: Elsevier. pp. 1–27.

Taylor, J. R. N., Schober, T. J. and Bean, S. R. 2006. Novel food and non-food uses for sorghum and millets. *Journal of Cereal Science* 44: 252–271.

Tovey, F. I. 1994. Diet and duodenal ulcer. *Journal of Gastroenterology & Hepatology* 9: 177–185.

Van Rensburg, S. J. 1981. Epidemiological and dietary evidence for a specific nutritional predisposition to esophageal cancer. *Journal of the National Cancer Institute* 67: 243–251.

Viswanath, V., Urooj, A. and Malleshi, N. G. 2009. Evaluation of antioxidant and antimicrobial properties of finger millet polyphenols (*Eleusine coracana*). *Food Chemistry* 114: 340–346.

Vivas, N. E., Waniska, R. D. and Rooney, L. W. 1992. Effects on proteins in sorghum, maize, and pearl millet when processed into acidic and basic to. *Cereal Chemistry* 69: 673–676.

Waniska, R. D. and Rooney, L. W. 2000. Structure and chemistry of the sorghum caryopsis. In: Smith, C. W. and Frederiksen, R. A., editors. *Sorghum: origin, history, technology, and production*. New York: Wiley. pp. 649–688.

Wankhede, D. B., Shehnaj, A. and Raghavendra Rao, M. R. 1979. Carbohydrate composition of finger millet (*Eleusine coracana*) and foxtail millet (*Setaria italica*). *Qualitas Plantarum Plant Foods for Human Nutrition* 28(4): 293–303.

Wu, Y., Lin, Q., Cui, T. and Xiao, H. 2014. Structural and physical properties of starches isolated from six varieties of millet grown in China. *International Journal of Food Properties* 17: 2344–2360.

Wu, Y., Wang, X., Xue, J. and Fan, E. 2017. Plant phenolics extraction from flos chrysanthemi: response surface methodology based optimization and the correlation between extracts and free radical scavenging activity. *Journal of Food Science* 82: 2726–2733.

Xiang, J., Apea-Bah, F. B., Ndolo, V. U., Katundu, M. C. and Beta, T. 2018. Profile of phenolic compounds and antioxidant activity of finger millet varieties. *Food Chemistry* 275: 361–368.

Xiao, Y., Zhang, Q., Miao, J., Rui, X., Li, T. and Dong, M. 2015. Antioxidant activity and DNA damage protection of mung beans processed by solid state fermentation with Cordyceps militaris SN-18. *Innovative Food Science & Emerging Technologies* 31: 216–225.

5

Millet Starch: Pasting, Rheological, and Morphological Properties

5.1 INTRODUCTION

India is a major grain producing country in the world, but changes in climatic conditions are the main issues for the production of grains and food security. Diverse types of food production are required; there is requirement for selecting the crops that can grow in drought and drastic conditions (Annor et al., 2017). Millets are the crops that can grow and survive in extreme conditions like low soil fertility, high alkalinity, and drought. Millet is a common term which is used for describing a range of small-seeded grains in two tribes Paniceae and Chlorideae of the family *Poaceae* (true grass). Millets are considered staple food for human, and they were around for 10,000 years ago before wheat and rice (Lu et al., 2009). Species of millets include browntop (*Brachiaria ramose*), barnyard (*Echinochloacrus galli*), finger (*Eleusine coaracana*), proso (*Panicum miliaceum*), kodo (*Paspalum scrobiculatum*), little (*Panicum sumatrense*), pearl (*Pennisetum glaucum*), and foxtail (*Setaria italica*). India is the leading producer (11,560,000 tons), followed by Niger (3,790,028 tons) (FAO, 2017).

Starch is the major reserve carbohydrate found in the plants and the major source of energy for humans. It is the most important part of different foods, and its characteristics and reaction with other food constituents, such as water and lipids, are most essential to the food industry and for human nutrition (Copeland et al., 2009). Millets are rich sources of starch, and it can be easily isolated from millets. Starch consists of two main components, i.e., amylose (linear) and amylopectin (branched). Amylose molecules are made up of α-(1–4)-linked D-glucose units, while amylopectin is made up α-(1–4)-linked and α-(1–6)-linked D-glucose linkages. The commercial importance of starches is due to its various functional properties, mainly its ability to change the texture of food products (Kaur et al., 2010).

The dynamic rheological tests for small deformation oscillatory measurements provide valuable data on viscoelastic characteristics of starch pastes without breaking the structure of starch molecules. These rheological measurements are vital for analyzing the molecular arrangement of starches (Gunasekaran & Mehmet Ak, 2000). Rheological properties of starch dispersions are similar to other food dispersions; these characteristics are considerably important for applications related with sensory analysis, understanding the structure, estimating textural properties, quality control, and processing of foods (Kim & Yoo, 2009).

There is growing interest in starch digestion kinetics due to the increasing prevalence of diseases such as diabetes and consumers' awareness of the relationship between food, nutrition, and health (Naidoo et al., 2015). The rate and extent of starch digestibility is nutritionally important, and it is usually determined by *in vitro* methods using pancreatic α-amylase containing amyloglucosidase to simulate the digestion in the small intestine (Englyst et al., 1992; McCleary & Monaghan, 2002). On the basis of rate and extent of digestibility, starch can be categorized into three types: starch that digests rapidly (rapidly digesting starch, RDS), a portion that digests slowly (slowly digesting starch, SDS), and a fraction that is resistant to digestion (resistant starch, RS) (Englyst et al., 1992).

Millets are an underutilized crop as compared to cereal and legumes; however, interest in their use has increased in recent years. The evaluation of the properties, structures, and applications of millet starch contributes greatly to the further development and industrialization of millet. Starch is mainly isolated from conventional source like potato and corn, but these crops are used as table food. But isolating starch from other sources likes millets can lead to decrease of the load on table food, so that this food may be utilized for those people who are suffering from starvation. This chapter will summarize the information about physicochemical, pasting, thermal, and rheological properties and *in vitro* digestibility of millet starches.

5.2 STARCH ISOLATION METHOD

Generally, millet starches are isolated using the wet milling method. For isolating the starches from millets, clean millet grains are steeped in aqueous solution and, after milling the grains, slurry is sieved to remove the fibrous matter. After sieving, centrifugation is done to separate the remaining protein and fibrous matter to purify the starch. Crude starch is repeatedly washed with water and centrifuged to remove the protein and fiber layer. Then starch is dried in a conventional oven at 40°C–50°C (Sandhu & Singh, 2005; Wu et al., 2014). Sharma et al. (2015) used alkali steeping method for extraction of starch from pearl millet flour. After steeping flour in 0.3% NaOH at 45 ± 2°C for 90 min, slurry is centrifuged to remove the nonstarch portion and starch is dried in a conventional oven. Babu et al. (2019) used the method reported by Dharmaraj et al. (2014) to isolate the starch from foxtail millet flour. Polished foxtail millet grains were steeped in deionized water for 15 h. The processes of grinding and sieving are repeated until the residue was free from starch content. Then this starch content is treated with NaOH, NaCl, and toluene solution to purify the starch content. Chao et al. (2014) used the alkali steeping method for isolation of proso millet starch. Wankhede et al. (1979) use toluene and mercuric chloride for extraction of starch from finger millet. Zhong et al. (2009) explained that pasting and rheological characteristics of starches are affected by the method used for extraction of starch. Variation in yield of starches from different sources may be due to variation in varieties, bonding of starch with other components, physiological condition of the seed, and the methods of evaluation (Wani et al., 2016). The starch isolation from millet grains is shown in Figure 5.1.

5.3 CHEMICAL COMPOSITION

Significant ($P < 0.05$) differences were observed in the chemical composition of millet starches (Table 5.1). Starch is semicrystalline in nature, with different levels of crystallinity. Amylopectin is mainly responsible for the crystallinity of starch granule, while the amorphous portions are associated with amylose (Zobel, 1988a, b). Amylose contents of 28.8%–31.9%, 15.64–19.46, 28.14%, 32.5%, and 3.96%–4.96% are observed for pearl millet starches (Hoover et al., 1996; Khatkar et al., 2013; Wu et al., 2014; Annor et al., 2014a; Suma & Urooj, 2015), 20%–27.1%, 13.22%, and 16.8%–26.8% for foxtail millet

100g millet seeds

↓

Grinding of seeds

↓

Steeping in NaOH (3g/L)
Solid liquid ratio (1:7) at 35°C for 28h

↓

Sieving (100mesh, 200mesh sieve)

↓

Centrifuged 3000×g (10min.)

↓

Re-suspension of settled starch layer in water

↓

Neutralized with HCl (0.1 mol/L)

↓

Drying 45°C

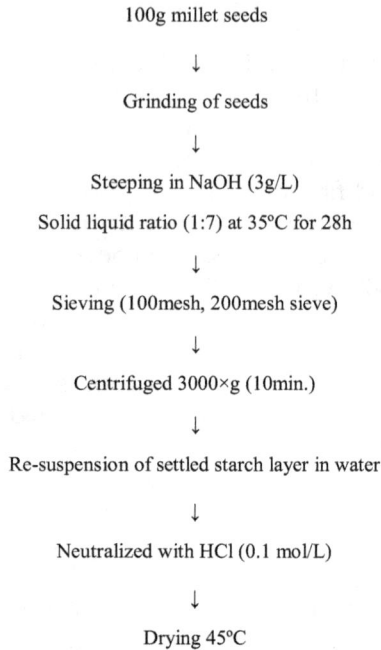

Figure 5.1 Separation of starch from grains (Chao et al., 2014).

(Fujita et al., 1996; Babu et al., 2019; Qi et al., 2019), 18.2%–18.5%, 38.6%, and 16.67%–18.90% for finger millet (Wankhede et al., 1979; Malleshi et al., 1986; Jayawardana et al., 2019), and 0.12%–0.25%, and 25.78%–29.08% for proso millet (Chao et al., 2014; Singh & Adedeji, 2017), respectively. Amylose content is a key factor for applications of starch because functional characteristics are affected by amylose content; a higher value of amylose content shows higher PT and lower PV and BV (Pycia et al., 2012). Wang and Copeland (2015) reported that amylose molecule is supposed to be primarily found in the core region of the starch granule and in amorphous regions between the crystalline lamellae; it is also observed that some amylose chains exist between the amylopectin clusters and these are thought to give structural strengthening. The proximate composition of starch can be affected by the method of starch isolation. Hoover (2001) stated that there is a need to defat the starch content; if this process is not used, amylose content is underestimated. Zhu (2016) stated that method of estimation of starch content affects the value of amylose content; different methods give different results for the same samples. Singh et al. (2006) reported that amylose content varies with starch source and climatic and soil conditions during the growth of grain. Physical properties of starches, *i.e.*, swelling power, gelatinization temperature, and crystalline structure, can be affected by amylose content (Cheetham & Tao, 1998). In 2017, Chen et al. reported a negative correlation between amylose content and swelling power. Naguleswaran et al. (2010) stated that amylose–lipid complexes and the arrangement of amylopectin molecules affect the solubility and swelling power of starches.

The main component of starches are amylose and amylopectin; other component also exist in the starch in lesser amounts, such as ash, fat, and protein content (Table 5.1). These minor components confirm the purity of starch; lesser values of these components indicates a purer starch. The lipids can form an inclusion complex with amylose or be trapped inside the granules. Starch isolation method affects the quantity of minor component in the isolated starch, and starch properties are influenced by these minor components.

Table 5.1 Starch Yield and Chemical Composition of Millet Starches

Sample	Starch Yield (%)	Amylose Content (%)	Protein (%)	Fat (%)	References
Pearl millet	42.7–47.5	13.6–18.1	0.32–0.75	0.27–0.46	Sandhu and Siroha (2017)
	34.50–39.40	3.96–4.96	0.53–0.55	0.37–0.38	Suma and Urooj (2015)
		28.14	0.31		Wu et al. (2014)
	60.2	22.8	0.68	0.92	Wankhede et al. (1990)
Finger millet	63.4	32.4			Annor et al. (2014b)
		28.40	0.58		Wu et al. (2014)
	55.7	38.6			Malleshi et al. (1986)
Foxtail millet		15.72	1.75	0.75	Babu et al. (2019)
		23.4			Li et al. (2019)
		16.8–26.8	0.10–0.26	0.21–0.51	Qi et al. (2019)
	67.5–68.7	3.3–11.4			Kim et al. (2009)

(Continued)

103

Table 5.1 (*Continued*) Starch Yield and Chemical Composition of Millet Starches

Sample	Starch Yield (%)	Amylose Content (%)	Protein (%)	Fat (%)	References
Proso millet	54.1	28.51	1.21	0.27	Singh and Adedeji (2017)
		0.12–20.57			Chao et al. (2014)
		0.45	0.38		Wu et al. (2014)
		29.2–32.6			Yanez et al. (1991)
Barnyard millet		27.68	0.44		Wu et al. (2014)
		20.2			Kumari and Thayumanavan (1998)

5.4 CHEMICAL STRUCTURE

Amylose and amylopectin are the two major polysaccharides of starch. Amylose is a linear structured chain in which glycosyl units are bonded with α-(1–4)-linkages, while amylopectin is a branched structure in which glycosyl units are connected with α-(1–6)-bonds (Bertoft, 2004). In 1992, Juliano classified starches on the basis of amylose content as waxy, 0%–2%; very low, 5%–12%; low, 12%–20%; intermediate, 20%–25%; and high 25%–33% amylose starches. Amylose is a small and linear molecule, whereas amylopectin is large branched molecule, and branches in amylopectin exist in a clustered manner. The tightly branched unit in the amylopectin is termed building block, and it has an internal chain length of 1–3 glucosyl residues (Pérez & Bertoft, 2010). Long-branch chains of amylopectin, like amylose, bind iodine to form a single helical complex during potentiometric titration, develop a blue color, and, consequently, increase the iodine affinity and the apparent amylose content of the starch (Jane et al., 1999). Jane and Chen (1992) stated that the amylopectin chain length distribution and amylose molecular size produce synergistic effects on the viscosity of starch pastes. Unit chain length distribution (CLD) is termed as the first structural level for starch molecular structure. It is classified into four groups based on previous divisions, namely, fa (degree of polymerization [DP] 6–12), fb1 (DP 13–25), fb2 (DP 25–36), and fb3 (DP > 36) (Hanashiro et al., 1996). The semicrystalline rings in starch molecules are made up of alternating amorphous and crystalline lamellae. The external unit chains of amylopectin tend to form the double helices, contributing to the formation of the crystalline lamellae. The internal part of amylopectin tends to form the amorphous lamellae in a clustered manner (Bertoft, 2013). Annor et al. (2014a) observed 32.5% amylose content for pearl millet starch, in which 15.8% were long chains and 16.7% were short chains, and the ratio of long chain to short chain was 0.95. Annor et al. (2014b) found average chain length, short chain length, long chain length, external chain length, and internal chain length of 18.0, 15.80, 52.12, 11.95, and 5.0, respectively, for pearl millet starch. Gaffa et al. (2004) also observed values of 13.2–13.8 and 5.8–6.2 for external and internal chain lengths for amylopectin of pearl millet starches.

5.5 PHYSICOCHEMICAL PROPERTIES

Swelling power and solubility provide information about the degree of interaction between starch chains within both the amorphous and crystalline zones (Singh et al., 2003). During heating of starch molecules in excess water, the crystalline structure of starch molecules get disrupted due to breakdown of hydrogen bond, and the exposed hydroxyl group of starch bonds with the water molecule due to which swelling and solubility power of starches increases (Wani et al., 2012). Important properties of starchy food such as pasting and rheological properties are affected by the swelling capacity of starches (Cruz et al., 2013). Ratnayake et al. (2002) reported that amylose content affects the solubility power of starches, while swelling power is influenced by amylopectin. Many factors influence the swelling power and solubility of starches such as lipid–amylose complexes, starch chains interactions in starch molecule, ratio of amylose and amylopectin, and molecular weight (Wani et al., 2010). In 2011, Nwokocha and Williams observed a negative correlation between swelling power and apparent amylose content and gelatinization temperature, but their correlation with crystallinity was not well explained. Swelling power of pearl millet starch was 14.8, 18.0–28.6; 14.1–17 (Wankhede et al., 1990; Hoover et al., 1996; Sandhu & Siroha, 2017); for foxtail millet, it was 13.0; 9.5 (Malleshi et al., 1986; Shaikh et al., 2015); finger millet, 11.2–11.6, 7.0 (Wankhede et al., 1979); and proso millet, 24.99 (Singh & Adedeji, 2017), respectively.

105

5.6 TRANSMITTANCE

Light transmittance provides information on the behavior of starch paste when the light passes through it. Paste clarity of starches is an essential characteristic of starch pastes, which is important to achieve the desired consistency for food products such as jellies and fruit pastes (Wu et al., 2014). Paste clarity is positively correlated with the particle size of starch granules; the larger the granule size, the more is the interference with dispersion, resulting more clarity of starch pastes (Singh et al., 2004). Achille et al. (2007) explained that paste clarity of starches depends on factors such as starch source, ratio of amylose/amylopectin, starch modifications, and addition of solutes. Jan et al. (2017) and Suriya et al. (2016) also explained that paste clarity is affected by chain lengths of amylose and amylopectin, granular swelling, and molecular structure as well as amount of leached amylose. The light transmittance of cooked starch paste of pearl millet starch was observed to be 1.80%–2.60% (Khatkar et al., 2013). Chao et al. (2014) compared light transmittance of waxy and nonwaxy proso millet starch and concluded that waxy starch (29.84%) had higher transmittance as compared to nonwaxy starch (21.57%). Impurity in starches, such as protein and lipids, is responsible for the lesser values of transmittance of starch pastes, which reduces the light transmittance by absorption and refraction (Craig et al., 1989). Khatkar et al. (2013) observed that retrogradation of the long chain of amylose and amylopectin occurs rapidly, which affects the light transmittance.

5.7 PASTING PROPERTIES

Pasting properties of starch can be calculated by using an amylograph, such as Brabender Amylograph and Rapid Visco-Analyzer (RVA) (Suh & Jane, 2003), or by using the starch cell of a dynamic rheometer (Sandhu & Siroha, 2017; Siroha et al., 2019a, b; Punia et al., 2019). RVA is the most common instrument used for measurement of viscosity of the sample; this analyzes the sample quickly and requires less amount of sample for analysis (Cozzolino, 2016). Pasting properties of millet starches are shown in Table 5.2 and Figure 5.2. Pasting properties provide information about the cooking characteristics of starches during heating and cooking cycles. Pasting of starch deals with changes that occur after gelatinization of starch, which includes swelling and leaching of starch components,

Table 5.2   Pasting Properties of Millet Starches

Instrument Used	Starch (%)	PV	BD	SV	Unit	PT (°C)	References
			Pearl millet				
Rheometer	8.0	1291–1853	146–518	506–961	mPa.s	74.0–85.4	Siroha et al. (2020)
Rheometer	16.6	4647–8303	2273–3344	1342–4722	mPa.s	72.4–73.9	Sandhu and Siroha (2017)
RVA	10.0	2826	1313	2439	mPa.s	77.3	Sharma et al. (2016)
BVA	10.0	300	137	124.5	BU	73.8	Shaikh et al. (2015)
RVA	6.0	1665–1998	414–769	8–502	cP	88.1–90.2	Khatkar et al. (2013)
			Foxtail millet				
RVA	7.6	867.1	179.1	442.2	cP	77.8	Li et al. (2019)
RVA	12.0	2203			cP		Babu et al. (2019)
RVA	14.0	4136–5100	2192–3529	1701–2259	cP	72.0–78.5	Liu et al. (2011)
BVA	10.0	485	40	415	BU	78.8	Kumari and Thayumanavan (1998)
			Finger millet				
BVA	7.0	280–330	20–30	110–130	BU	80.5–86.5	Madhusudhan and Tharanathan (1995)
BVA	9.5					67.5	Wankhede et al. (1979)
			Proso millet				
Rheometer	14.0	4.60	2.60	1.69	Pa.s	79.23	Singh and Adedeji (2017)
RVA	8.0	2059–3515	440–967	197–1161	cP	63.60–63.80	Chao et al. (2014)
RVA	10.7	2822	1854	501	cP	76.0	Sun et al. (2014)

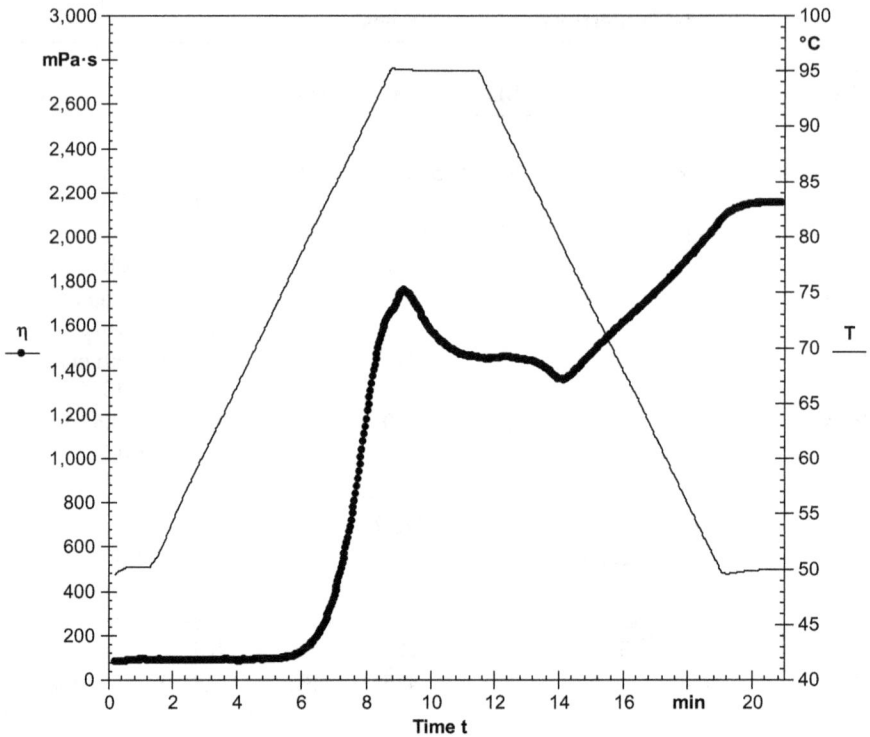

Figure 5.2 Pasting properties of millet starch.

and due to applied force starch paste viscosity increases (Atwell et al., 1988; Tester & Morrison, 1990). Ghiasi et al. (1982) explained that during heating of starch at high temperature, amylose content leaches out from the starch granule, which results in increase in viscosity. Peak viscosity (PV) of starch is the maximum viscosity obtained by the starch granule during heating without breakdown of the starch granule (Jan et al., 2016). PV of millet starches has been observed by many researchers, PV for pearl millet was found to be 1665–1998, 2826, and 4647–8303 mPa.s (Khatkar et al., 2013; Sharma et al., 2016; Sandhu & Siroha, 2017); for foxtail millet, 745–1855 RVU, 775–1117 cP (Kim et al., 2009; Li et al., 2019); for finger millet, 1580–1720 BU, 280–330 RVU (Wankhede et al., 1979; Madhusudhan & Tharanathan, 1995); and for proso millet, 2822 cP, 4.60 Pa.s (Sun et al., 2014;

Singh & Adedeji, 2017). Breakdown viscosity (BV) deals with the resistance of starch paste during shear and heat, while retrogradation capacity of starch is measured by setback viscosity (SV). Asante et al. (2013) explained that starches that have higher value for BV and low value for SV showed good cooking quality. Pasting temperature (PT) is the temperature at which viscosity of the starch paste starts to increase during heating of starch; PT provides information on the temperature required to cook the starch (Naivikul & Appolonia, 1979). Tannin and phenolic content affect the pasting properties of starches (Beta & Corke, 2004; Barros et al., 2012). The pasting characteristics of starch are an efficient method for describing starch functionality with its structural features and the potential industrial application in food products dependent on the viscosity and thickening behavior of starch (Cruz et al., 2013).

5.8 THERMAL PROPERTIES

Thermal properties of millet starches are reported in Table 5.3. Starch granules when heated in excess of water undergo significant structural and morphological changes, including starch swelling due to water absorption, loss of crystallinity due to the amylopectin double helix dissociation, and amylose leaching to the water phase (Cruz et al., 2013). Thermal properties of millets starches are shown in Table 5.3. These sets of changes are generally referred to as starch gelatinization. Gelatinization temperatures are important for selecting ideal varieties with specified physicochemical properties of starches and for various food application requirements (Morales-Martínez et al., 2014). Chavan et al. (1999) reported that gelatinization involves the uncoiling and melting of the external chains of amylopectin that are packed together as double helices in clusters. Many techniques are used to determine the gelatinization temperature of starch, such as differential scanning calorimetry (DSC) (Biliaderis et al., 1980), polarized light microscopy equipped with a hot stage (Li et al., 2013), thermomechanical analysis (Biliaderis et al., 1986), and nuclear magnetic resonance spectroscopy (Gonera & Cornillon, 2002), with other methods for the degree of starch gelatinization, such as X-ray scattering (Jenkins et al., 1994) and Fourier transform infrared spectroscopy (Rubens & Heremans, 2000). Crystallite quality and the crystallinity of the starch are analyzed by the peak temperature (T_p) and the enthalpy of gelatinization (ΔHg), respectively (Tester & Morrison, 1990). Onset temperature (T_o) and conclusion temperature (T_c)

109

Table 5.3 Thermal properties of millet starches

Starch:water ratio (w/w)	Heating rate (°C/min)	Gelatinization parameters				References
		T_o (°C)	Tp (°C)	Tc (°C)	ΔHgel (J/g)	
		Pearl millet				
1:2	5	63.4–67.7	69.3–71.6	74.5–76.3	10.6–12.4	Sandhu and Siroha (2017)
1:4	10	59.9	62.6	73.6	11.9	Sharma et al. (2016)
1:3	10	69.6	73.7	87.2	9.8	Shaikh et al. (2015)
1:2	10	64.7	69.7	74.6	3.5	Wu et al. (2014)
1:3	5	62.8	67.9		12.3	Annor et al. (2014a)
		Foxtail millet				
		62.5–75.1			8.2–13.5	Fujita et al. (1996)
1:3	10	66.4–69.6	71.0–74.2	74.1–79.5	−0.44 to 8.21	Kim et al. (2009)
1:3	10	65.4–68.0	70.9–73.2	75.6–80.3	11.8–13.4	Qi et al. (2019)
1:3	10	68.1	72.8	77.2	21.0	Li et al. (2019)
		Finger millet				
1:2	10	63.41	68.64	74.0	2.93	Wu et al. (2014)
		Proso millet				
1:3	10	67.9–73.5	74.6–77.1	80.4–81.3	10.37–13.06	Chao et al. (2014)
1:2	10	68.65	71.37	80.4	16.10	Sun et al. (2014)
1:2	10	72.93	78.61	94.55	3.83	Singh and Adedeji (2017)

provide data on the boundaries of the different phases in a semicrystalline material like starch (Biliaderis et al., 1986). In 1996, Noda et al. explained that DSC parameters (T_o, T_p, T_c, and ΔHgel) are affected by molecular design of the crystalline region, which corresponds to the distribution of amylopectin short chains, and not by the proportion of the crystalline region, which corresponds to amylose/amylopectin ratio. Sang et al. (2008) explained that T_p is an indicator of crystallite quality related to the double helix length, whereas ΔH a measure of the loss of molecular order. Cooke and Gidley (1992) reported that ΔH shows the loss of double helical order rather than the loss of crystallinity. Krueger et al. (1987) reported that the higher the amylopectin content of the starch, the narrower was the temperature range of gelatinization. Knutson et al. (1982) also found that more heterogeneous granules broaden the enthalpy range. High transition temperatures have been reported to result from a high degree of crystallinity, due to which structural stability and delay of gelatinization occurs (Barichello et al., 1990). A direct relationship is observed between the amylose content of the starches and their gelatinization temperature. Lawal et al. (2011) stated that maximum amorphous region of starch is made up of amylose component, while amylopectin is responsible for crystalline region.

5.9 RETROGRADATION

Starch retrogradation is a process in which reassociation of amylose and amylopectin chains of gelatinized starch occurs and a more ordered structure is formed. During retrogradation, a double helix structure is formed by amylose bonding with other glucose units, where recrystallization of amylopectin occurs through association of small chains (Singh et al., 2003). Wang et al. (2015) stated that during gelatinization and retrogradation, changes occurs in the starch structure that are important determinants of its functional properties for food processing, during digestion, and in industrial applications. Adebowale et al. (2005) reported retrogradation properties (40°C for 24 h) for finger millet starch and observed value for transitions temperature (T_o, T_p, and T_c) and ΔH of 54.65°C, 58.06°C, 62.32°C, and 1.62 J/g. Chao et al. (2014) evaluated the retrogradation capacity of proso millet starch at different times (4, 8, and 32 h) and observed average retrogradation percentage reaching 10.02%. Shaikh et al. (2017) observed T_o, T_p, T_c, and ΔH values of 48.18°C, 58.44°C, 72.14°C, and 4.84 (J/g) for retrogradation of pearl millet starch. Retrogradation enthalpy, ΔH_R, shows the unraveling and melting of double helices formed during retrogradation, which is a function of amylopectin unit chain length distribution (Lu et al., 1997).

5.10 RHEOLOGICAL PROPERTIES

Rheological methods are used by many researchers to determine the viscoelasticity characteristics of the starch pastes (Sandhu & Siroha, 2017; Siroha & Sandhu, 2018; Punia et al., 2019). Different methods have been used to characterize the rheological properties of a starch gel (Figure 5.3). The most frequently used methods include: (1) determination of starch gel strength using a texture analyzer, which provides a "single-point"

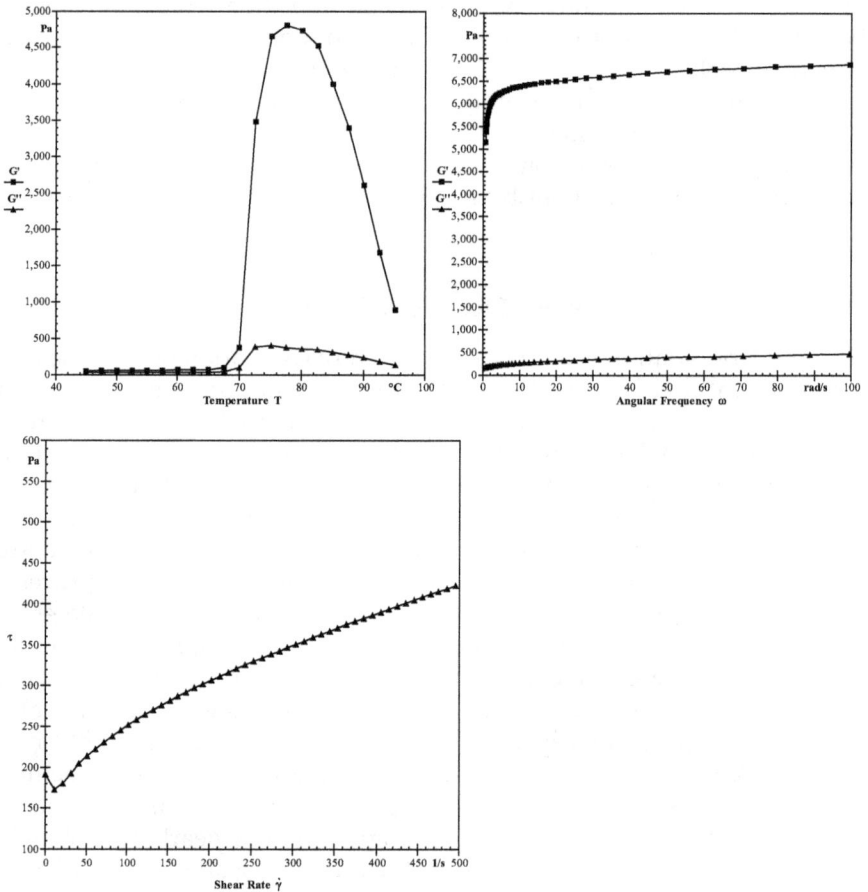

Figure 5.3 Rheological properties of millet starch.

measurement and (2) dynamic modulus analysis of starch gel using a dynamic rheometer, which allows continuous assessment of starch gel at various temperatures and shear rates (Ai & Jane, 2015). The most common method of studying the viscoelastic properties of starch is by using a dynamic rheometer. The continuous assessment of the starch suspensions during temperature and frequency sweep tests can be done with a dynamic rheometer. Rheological properties of millet starches are observed by various researchers (Wu et al., 2014; Sharma et al., 2015; Sandhu & Siroha, 2017; Siroha et al., 2019b). G′ (storage modulus) is a measure of the energy stored in the material and recovered from it per cycle, while G″ (loss modulus) is a measure of the energy dissipated or lost per cycle of sinusoidal deformation (Ferry, 1980). It is reported that amylose content affects the rate of retrogradation and viscoelastic properties of starches; starches with high amylose content had a higher value for G′ for cooked starch pastes (Singh et al., 2011). Many applications of polymers depend upon their viscoelastic properties. The swelling of starch granules to fill all the available volume in a system is related with the formation of a three-dimensional network of swollen granules that occurs due to intergranular contact required for the initial increase in G′ (Eliasson, 1986). When starch is heated in a rheometer, initially G′ value starts to increase and reaches the maximum value during heating; after attaining peak value due to breakdown of starch granule structure after further heating, a decrease in G′ value is observed. G′ and G″ values for pearl millet starches were 3613–6334 and 352–441 Pa, respectively (Sandhu & Siroha, 2017), and G′ and G″ values were 361–922 and 62.4–89.9 Pa, respectively, during heating of starch suspension (Siroha et al., 2020). Ai and Jane (2015) explained that chemical structure of starch, starch concentration, pasting conditions (e.g., temperature, shear rate, and heating rate), and storage conditions (temperature and time) affect the rheological properties of starch paste and gel. tanδ is used for determination of the viscoelastic behavior of starch pastes; a tanδ value less than 1 indicates elastic behavior, while a tanδ value more than 1 shows viscous behavior of starch pastes.

The frequency sweep test provides significant information about the gel structure of starch pastes (Kaur et al., 2008). In addition to G′, G″ and tanδ (G″/G′) were observed to reflect the dynamic elastic nature of the gels, indicating the relative measure of the associated energy loss versus the energy stored per deformation cycle (Toker et al., 2013). Different constitutive equations are employed to model the mechanical behavior of deformable particles. In particular, three elastic constitutive equations are considered in the literature, namely Power law, Casson model, and

Herschel–Bulkey model. Steady shear properties of millet starches are studied by many researchers, and yield stress (σ_0), consistency index (K), and flow behavior index (n) are studied for various millet cultivars (Sharma et al., 2016; Sandhu & Siroha, 2017; Siroha et al., 2020). Yield stress indicative of the minimum force required to initiate flow of starch paste, for different cultivars ranged from 25.6 to 183.4 and 40.73 to 115.72 Pa (Sandhu & Siroha, 2017; Siroha et al., 2020). Differences in yield stress values may be due to the differences in the concentration used for sample analysis. K value of pearl millet starches ranged from 0.729 to 3.998 Pa.s. (Siroha et al., 2020), while Sandhu and Siroha (2017) observed 1.69–54.5 Pa.s. values for millet starches. The difference between K values may be due to the concentration of starch used for measurement. To analyze the fluids and semi fluid nature of pastes, n value is commonly used; behavior of pastes with n value of 1 indicates a Newtonian fluid, n value of less than 1 shows shear thinning, and n value of greater than 1 indicates a shear thinking fluid behavior. Bhandari et al. (2002) reported that due to high shear rate breakdown of the intra- and intermolecular associative bonding system in starch network, micelles occur, which results in shear thinning behavior.

5.11 X-RAY DIFFRACTION

X-ray diffractometry has been widely used to analyze the properties of the crystalline structure of starch granules (Zobel et al., 1988). Starch granules had a semi-crystalline structure related to different polymorphic forms and, based on this, starch can be differentiated into three types, namely A, B, and C (Buléon et al., 1998). A-type X-ray diffractogram (XRD) crystalline pattern is mostly characterized with 2θ values of about 15.3, 17.8, and 23.2; B type starches have 2θ values of about 5.5, 15, 17, 22.2, and 24 (Jayakody et al., 2005). C type starches are a combination of A and B types and typical for legume starches (Donald, 2004). Millet starches showed A-type XRD pattern (Kim et al., 2012; Sun et al., 2014; Sandhu & Siroha, 2017; Qi et al., 2019). Perez et al. (2009) reported that gelatinization temperature of A-type polymorphs starches generally increase with increase in the crystallinity of starches while the reverse is observed for B-type polymorphs. There are many factors such as growth conditions, harvesting time, biological origin, and amylose and amylopectin contents, which are responsible for variation in the diffraction patterns of starches (Zhou et al., 2010). Benmoussa et al. (2006) stated that digestibility of starch in animals is affected by degree of crystallinity.

The degree of crystallinity is influenced by the molecular structure of amylopectin (unit chain length, extent of branching, molecular weight, and polydispersity) (Bao, 2004). Liu (2005) explained the effect of amylose content on crystallinity; amylose content has little effect on the crystallinity of A-type starches; however, lesser values for crystallinity are observed for higher amylose-content starches but there is no such trend is observed for C-type starches for crystallinity (Liu, 2005). The crystallinity of starches is related to the packing of the amylopectin double helices, while the amorphous region is associated with amylose (Zobel, 1988a,b).

5.12 MORPHOLOGICAL PROPERTIES

Light microscopy and scanning electron microscopy (SEM) can be used to study the morphological properties of starch granules (Fannon et al., 1992; Jane et al., 1994) (Figure 5.4 & Table 5.4). SEM is used to study the granule morphology more accurately than light microscopy. Chmelik et al. (2001) reported that high resolution of SEM also provides a more extensive

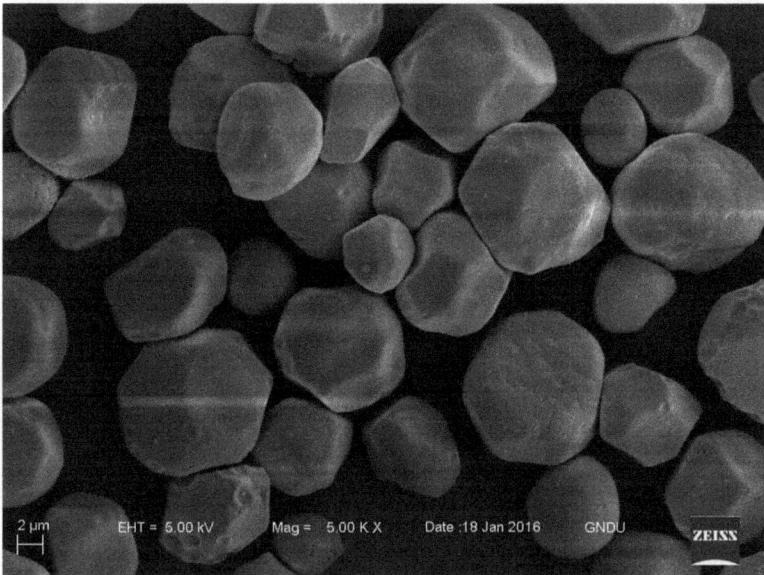

Figure 5.4 Scanning electron micrograph of millet starch.

Table 5.4 Morphology of Starch Granules

Shape of Starch Granules	Size (μm)	References
	Pearl millet	
Small to large, spherical and polygonal		Sandhu and Siroha (2017)
Polygonal shape with a few spherical-shaped granules with indentations and creases	6–12	Shaikh et al. (2015)
Large polygonal and few small spherical granules	3–15	Wu et al. (2014)
Polygonal with a few spherical granules	3.5–23	Annor et al. (2014a)
	Foxtail millet	
Polygonal shape	6.8–11.8	Fujita et al. (1996)
Polygonal and rarely elliptical	10.1–25.0 and 4.7–12.5	Kim et al. (2009)
	Finger millet	
Spherical, polygonal, and rhombic		Malleshi et al. (1986)
Polygonal		Annor et al. (2014a)
Polygonal	3–10	Wu et al. (2014)
	Proso millet	
Polygonal granules with indentations	1·8–13·5	Yanez et al. (1991)
Polygonal granules	5–12	Sun et al. (2014)
Polygonal-shaped with smooth edges and spherical	6.12–6.44	Chao et al. (2014)

approach on granule surface properties and granule morphology. Starch from different botanical sources presents characteristic shapes, sizes, and morphologies (Jane et al., 1994; Lindeboom et al., 2004). Lindeboom et al. (2004) explain the various applications of small granule size starches such as fat replacers, carrier material, fillers for biodegradable films, and cosmetic ingredients. Fannon et al. (1993) hypothesized these pores of sorghum and millet starches as the openings to channels that provide access to the granule interior. Physicochemical, functional, and nutritional properties of starches are affected by the shape and size of starch granules; larger granules are responsible for high paste viscosity, while small granules had higher digestibility (Bello-Pérez et al., 2010). Sandhu and Siroha (2017) reported starch granules varying in size and shape from small to large and spherical and polygonal for pearl millet starch. A similar granular shape was observed by Annor et al. (2014a) and Chao et al. (2014).

5.13 DIGESTIBILITY PROPERTIES

The glycemic index (GI) is a method for ranking foods with respect to their blood glucose-raising potential (Kaur & Sandhu, 2010). GI is most appropriately used to compare foods within a category of foods. Starch is classified into rapidly digestible starch (RDS), slowly digestible starch (SDS), and resistant starch (RS) on the basis of the rate and extent of its digestion. RDS is digested in vitro within 20 min, SDS is digested between 20 and 120 min, and RS is the starch not hydrolyzed after 120 min of incubation (Englyst et al., 1992). Guraya et al. (2001) reported that granule size of starch affects the starch digestibility; a larger granule size shows lesser values for digestibility. Digestibility of A-type starch is generally higher as compared to B-type starch, and C-type starches have an intermediate digestion rate between A- and B-type starches (Biliaderis, 1991; Zhao & Gu, 2007). Differences in the digestibility of starches among different species have been related to the interplay of many factors, such as source of starch (Ring et al., 1988), size of starch granules (Lindeboom et al., 2004), degree of crystallinity (Hoover & Sosulski, 1985), and amylose/amylopectin ratio (Hoover & Sosulski, 1991). Some other factors also affect the starch digestibility such as amylose/amylopectin ratio, granule size, amylose–lipid complexes, and molecular structure of amylopectin (Chung et al., 2008, 2011). Sandhu and Siroha (2017) observed RDS, SDS, and RS contents of 46.3%–50.1%, 37.2%–38.7%, and 9.7%–16.5%, respectively, for pearl millet starches. Qi et al. (2019) reported RDS, SDS, and RS contents of 35.7%–39.1%, 45.7%–53.0%, and 11.3%–15.7%, respectively, for foxtail millet starches. Jayawardana et al. (2019) observed RDS, SDS, and RS contents of 10.90%–1197%, 43.38%–49.15%, and 3.75%–4.58%, respectively, for finger millet starch.

117

5.14 APPLICATIONS

Millet starches can be used in different food products. Like other starches, it can be used as a thickener, gelling agent, and bulking agent various foods. Native starch has many limitations to overcome; these limitations are generally modified using physical, chemical, and enzymatic methods. Succinylated starches are used for various food products such as oil-in-water emulsions for food, pharmaceutical, and industrial products. Babu and Mohan (2019) succinylated foxtail millet starch, which can be used in these food products. Succinylated starches also can be used as fat replacers for various foods. Plastic is used as a packaging material for various products, which is an environmental issue. Starch can be used for making biodegradable films to solve this problem. Sandhu et al. (2020) reported starch films from pearl millet starch and carrageenan gum. Many researchers reported HMT modification of millet starches (Adebowale et al., 2005; Sharma et al., 2015; Sandhu et al., 2020a). HMT modification improves the RS content, which has many health benefits. It can be used in food products for enhancing nutritional value. Millet starch can be used in the pharmaceutical industry as well. Odeku and Alabi (2007) reported utilization of pearl millet starch in tablet formulation. Jhan et al. (2020) prepare the nanostarch from pearl millet starch. Nanostarch has many applications in food as well as pharmaceutical industries. Nanostarch can be used for as a binder for vitamin, to provide flavor, for pigment etc. Nanostarch also be used as a drug delivery system. These starches release the drug slowly, thus prolonging the efficacy.

5.15 CONCLUSION

In this chapter, physicochemical, pasting, rheological, and in vitro digestibility properties have been discussed, and diversity in these properties is found for different cultivars. Millet starches show A-type crystalline pattern, which is mainly found in the cereal starches. Granule size of millet starches is observed to be small to large, with spherical and polygonal shapes. Higher content of RS is observed for millet starches, which show many health benefits. Rheological properties of millet starches showed shear thinning and viscoelastic behavior for starch paste. A wide range of different functional properties is observed for millet starches due to which the scope of use of millet starches in food and other industries is increased. Further research may be conducted on utilization of millet starches in different food products and their modification by chemical or enzymatic methods to make it more useful for food processors.

REFERENCES

Achille, T. F., Nrsquo, A., Georges, G. and Alphonse, K. 2007. Contribution to light transmittance modelling in starch media. *African Journal of Biotechnology* 6(5): 569–575.

Adebowale, K. O., Afolabi, T. A. and Olu-Owolabi, B. I. 2005. Hydrothermal treatments of Finger millet (*Eleusine coracana*) starch. *Food Hydrocolloids* 19(6): 974–983.

Ai, Y. and Jane, J. L. 2015. Gelatinization and rheological properties of starch. *Starch/Stärke* 67: 213–224.

Annor, G. A., Marcone, M., Bertoft, E. and Seetharaman, K. 2014a. Physical and molecular characterization of millet starches. *Cereal Chemistry* 91: 286–292.

Annor, G. A., Marcone, M., Bertoft, E. and Seetharaman, K. 2014b. Unit and internal chain profile of millet amylopectin. *Cereal Chemistry* 91: 29–34.

Annor, G. A., Tyl, C., Marcone, M., Ragaee, S. and Marti, A. 2017. Why do millets have slower starch and protein digestibility than other cereals? *Trends in Food Science & Technology* 66: 73–83.

Asante, M. D., Offei, S. K., Gracen, V., Adu-Dapaah, H., Danquah, E. Y., Bryant, R. and McClung, A. 2013. Starch physicochemical properties of rice accessions and their association with molecular markers. *Starch/Stärke* 65(11–12): 1022–1028.

Atwell, W. A., Hood, L. F., Lineback, D. R., Varriano-Marston, E. and Zobel, H. F. 1988. The terminology and methodology associated with basic starch phenomena. *Cereal Foods World* 33: 306–311.

Babu, A. S. and Mohan, R. J. 2019. Influence of prior pre-treatments on molecular structure and digestibility of succinylated foxtail millet starch. *Food Chemistry* 295: 147–155.

Babu, A. S., Mohan, R. J. and Parimalavalli, R. 2019. Effect of single and dual-modifications on stability and structural characteristics of foxtail millet starch. *Food Chemistry* 271: 457–465.

Bao, J. 2004. The functionality of rice starch. In: Eliasson, A.-C., editor. *Starch in food. Structure, function and application*. UK: Woodhead Publishing. pp. 258–289.

Barichello, V., Yada, R. Y., Coffin, R. H. and Stanley, D. W. 1990. Low temperature sweetening in susceptible and resistant potatoes: starch structure and composition. *Journal of Food Science* 54: 1054–1059.

Barros, F., Awika, J. M. and Rooney, L. W. 2012. Interaction of tannins and other sorghum phenolic compounds with starch and effects on in vitro starch digestibility. *Journal of Agricultural and Food Chemistry* 60: 11609–11617.

Bello-Pérez, L. A., Sánchez-Rivera, M. M., Núñez-Santiago, C., Rodríguez-Ambriz, S. L. and Román-Gutierrez, A. D. 2010. Effect of the pearled in the isolation and the morphological, physicochemical and rheological characteristics of barley starch. *Carbohydrate Polymers* 81: 63–69.

Benmoussa, M., Suhendra, B., Aboubacar, A. and Hamaker, B. R. 2006. Distinctive sorghum starch granule morphologies appear to improve raw starch digestibility. *Starch/Stärke* 58: 92–99.

Bertoft, E. 2004. Analysing starch structure. In: Eliasson, A. C., editor. *Starch in food–structure, function, and application.* Cambridge: Woodhead Publishing. pp. 57–71.

Bertoft, E. 2013. On the building block and backbone concepts of amylopectin structure. *Cereal Chemistry* 90: 294–311.

Beta, T. and Corke, H. 2004. Effect of ferulic acid and catechin on sorghum and maize starch pasting properties. *Cereal Chemistry* 81: 418–422.

Bhandari, P. N., Singhal, R. S. and Kale, D. D. 2002. Effect of succinylation on the rheological profile of starch pastes. *Carbohydrate Polymers* 47: 365–371.

Biliaderis, C. G. 1991. The structure and interactions of starch with food constituents. *Canadian Journal of Physiology & Pharmacology* 69: 60–78.

Biliaderis, C. G., Maurice, T. J. and Vose, J. R. 1980. Starch gelatinization phenomena studied by differential scanning calorimetry. *Journal of Food Science* 45: 1669–1674.

Biliaderis, C. G., Page, C. M., Maurice, T. J. and Juliano, B. O. 1986. Thermal characterization of rice starches: a polymeric approach to phase transitions of granular starch. *Journal of Agricultural and Food Chemistry* 34: 6–14.

Buléon, A., Colonna, P., Planchot, V. and Ball, S. 1998. Starch granules: structure and biosynthesis. *International Journal of Biological Macromolecules* 23: 85–112.

Chao, G., Gao, J., Liu, R., Wang, L., Li, C., et al. 2014. Starch physicochemical properties of waxy proso millet (*Panicum Miliaceum* L.). *Starch/Stärke* 66: 1005–1012.

Chavan, U. D., Shahidi, F., Hoover, R. and Perera, C. 1999. Characterization of beach pea (*Lathyrus maritimus* L.) starch. *Food Chemistry* 65: 61–70.

Cheetham, N. W. and Tao, L. 1998. Variation in crystalline type with amylose content in maize starch granules: an X-ray powder diffraction study. *Carbohydrate Polymers* 36: 277–284.

Chen, X., Li, X., Mao, X., Huang, H., Wang, T., et al. 2017. Effects of drying processes on starch-related physicochemical properties, bioactive components and antioxidant properties of yam flours. *Food Chemistry* 224: 224–232.

Chmelik, J., Krumlová, A., Budinská, M., Kruml, T., Psota, V., et al. 2001. Comparison of size characterization of barley starch granules determined by electron and optical microscopy, low angle laser light scattering and gravitational field-flow fractionation. *Journal of the Institute of Brewing* 107: 11–17.

Chung, H.-J., Lim, H. S. and Lim, S. T. 2008. Effect of partial gelatinization and retrogradation on the enzymatic digestion of waxy rice starch. *Journal of Cereal Science* 43: 353–359.

Chung, H. J., Liu, Q., Lee, L. and Wie, D. 2011. Relationship between the structure, physicochemical properties and in vitro digestibility of rice starches with different amylose contents. *Food Hydrocolloids* 25: 968–975.

Cooke, D. and Gidley, M. J. 1992. Loss of crystalline and molecular order during starch gelatinization. Origin of the enthalpic transition. *Carbohydrate Research* 227: 103–112.

Copeland, L., Blazek, J., Salman, H. and Tang, M. C. 2009. Form and functionality of starch. *Food Hydrocolloids* 23: 1527–1534.

Cozzolino, D. 2016. The use of the rapid visco analyser (RVA) in breeding and selection of cereals. *Journal of Cereal Science* 70: 282–290.

Craig, S. A. S., Maningat, C. C., Seib, P. A. and Hoseney, R. C. 1989. Starch paste clarity. *Cereal Chemistry* 66: 173–182.

Cruz, B. R., Abraão, A. S., Lemos, A. M. and Nunes, F. M. 2013. Chemical composition and functional properties of native chestnut starch (*Castanea sativa Mill*). *Carbohydrate Polymers* 94: 594–602.

Dharmaraj, U., Parameswara, P., Somashekar, R. and Malleshi, N. G. 2014. Effect of processing on the microstructure of finger millet by X-ray diffraction and scanning electron microscopy. *Journal of Food Science & Technology* 51: 494–502.

Donald, A. M. 2004 In: Eliasson, A. C., editor. *Starch in food: structure, function and applications.* Cambridge/New York: Woodhead Publishing/CRC Press. pp. 156–184.

Eliasson, A. C. 1986. Viscoelastic behaviour during the gelatinization of starch I. Comparison of wheat, maize, potato and waxy-barley starches. *Journal of Texture Studies* 17: 253–265.

Englyst, H. N., Kingman, S. M. and Cummings, J. H. 1992. Classification and measurement of nutritionally important starch fractions. *European Journal of Clinical Nutrition* 45: S33–S50.

Fannon, J. E., Hauber, R. J. and BeMiller, J. N. 1992. Surface pores of starch granules. *Cereal Chemistry* 69: 284–288.

Fannon, J. E., Shull, J. M. and BeMiller, J. N. 1993. Interior channels of starch granules. *Cereal Chemistry* 70: 611–613.

FAO (Food and Agricultural Organisation of the United Nations) 2017. http://www.fao.org/faostat/en/#data/QC (Accessed 15th June 2019).

Ferry, J. D. 1980. *Viscoelastic properties of polymers.* 3rd ed. New York: J. Wiley and Sons, (Chapter 11).

Fujita, S., Sugimoto, Y., Yamashitab, Y. and Fuwa, H. 1996. Physicochemical studies of starch from millet (*Setaria italica Beauv.*). *Food Chemistry* 55: 209–213.

Gaffa, T., Yoshimoto, Y., Hanashiro, I., Honda, O., Kawasaki, S. and Takeda, Y. 2004. Physicochemical properties and molecular structures of starches from millet (*Pennisetum typhoides*) and sorghum (*Sorghum bicolor L. moench*) cultivars in Nigeria. *Cereal Chemistry* 81: 255–260.

Ghiasi, K., Marston, V. K. and Hoseney, R. C. 1982. Gelatinization of wheat starch. II starche-surfactant interaction. *Cereal Chemistry* 59: 86–88.

121

Gonera, A. and Cornillon, P. 2002. Gelatinization of starch/gum/sugar systems studied by using DSC, NMR, and CSLM. *Starch/Stärke* 54: 508–516.

Gunasekaran, S. and Mehmet Ak, M. 2000. Dynamic oscillatory shear testing of foods-selected applications. *Trends in Food Science & Technology* 11: 115–127.

Guraya, H. S., James, C. and Champagne, E. T. 2001. Effect of cooling, and freezing on the digestibility of debranched rice starch and physical properties of the resulting material. *Starch/Stärke* 53: 64–74.

Hanashiro, I., Abe, J. and Hizukuri, S. 1996. A periodic distribution of the chain length of amylopectin as revealed by high-performance anion-exchange chromatography. *Carbohydrate Research* 283: 151–159.

Hoover, R. 2001. Composition, molecular structure, and physico-chemical properties of tuber and root starches: a review. *Carbohydrate Polymers* 45: 253–267.

Hoover, R., and Sosulski, F. 1991. Composition, structure, functionality and chemical modification of legume starches: A review. *Canadian Journal of Physiology and Pharmacology* 69: 79–92.

Hoover, R. and Sosulski, F. 1985. Studies on the functional characteristics and digestibility of starches from *Phaseolus vulgaris* biotypes. *Starch/Stärke* 37: 181–191.

Hoover, R., Swamidas, G., Kok, L. S. and Vasanthan, T. 1996. Composition and physicochemical properties of starch from pearl millet grains. *Food chemistry* 56: 355–367.

Jan, K. N., Panesar, P. S., Rana, J. C. and Singh, S. 2017. Structural, thermal and rheological properties of starches isolated from Indian quinoa varieties. *International Journal of Biological Macromolecules* 102: 315–322.

Jan, R., Saxena, D. C. and Singh, S. 2016. Pasting, thermal, morphological, rheological and structural characteristics of Chenopodium (*Chenopodium album*) starch. *LWT-Food Science & Technology* 66: 267–274.

Jane, J. L. and Chen, J. F. 1992. Effect of amylose molecular size and amylopectin branch chain length on paste properties of starch. *Cereal Chemistry* 69(1): 60–65.

Jane, J., Chen, Y. Y., Lee, L. F., McPherson, A. E., Wong, K. S., Radosavljevic, M. and Kasemsuwan, T. 1999. Effects of amylopectin branch chain length and amylose content on the gelatinization and pasting properties of starch. *Cereal Chemistry* 76: 629–637.

Jane, L., Kasemsuwan, T., Leas, S., Zobel, H. and Robyt, J. F. 1994. Anthology of starch granule morphology by scanning electron microscopy. *Starch/Stärke* 46: 121–129.

Jayakody, L., Hoover, R., Liu, Q. and Weber, E. 2005. Studies on tuber and root starches. I. Structure and physicochemical properties of innala (*Solenostemon rotundifolius*) starches grown in Sri Lanka. *Food Research International* 38: 615–629.

Jayawardana, S. A. S., Samarasekera, J. K. R. R., Hettiarachchi, G. H. C. M., Gooneratne, J., Mazumdar, S. D. and Banerjee, R. 2019. Dietary fibers, starch fractions and nutritional composition of finger millet varieties cultivated in Sri Lanka. *Journal of Food Composition and Analysis* 82: 103249.

Jenkins, P. J., Comerson, R. E., Donald, A. M., Bras, W., Derbyshire, G. E., Mant, G. R. and Ryan, A. J. 1994. In situ simultaneous small and wide angle X-ray scattering: a new technique to study starch gelatinization. *Journal of Polymer Science Part B: Polymer Physics* 32: 1579–1583.

Jhan, F., Shah, A., Gani, A., Ahmad, M. and Noor, N. 2020. Nano-reduction of starch from underutilised millets: effect on structural, thermal, morphological and nutraceutical properties. *International Journal of Biological Macromolecules* 159: 1113–1121.

Juliano, B. O. 1992. Structure, chemistry, and function of the rice grain and its fractions. *Cereal Food World* 37: 772–774.

Kaur, M. and Sandhu, K. S. 2010. *In vitro* digestibility, structural and functional properties of starch from pigeon pea (Cajanus cajan) cultivars grown in India. *Food Research International* 43: 263–268.

Kaur, M., Sandhu, K. S. and Lim, S. 2010. Microstructure, physicochemical properties and in vitro digestibility of starches from different Indian lentil (*Lens culinaris*) cultivars. *Carbohydrate Polymers* 79: 349–355.

Kaur, L., Singh, J., Singh, H. and McCarthy, O. J. 2008. Starch–cassia gum interactions: a microstructure-rheology study. *Food Chemistry* 111(1): 1–10.

Khatkar, B. S., Rajneesh, B. and Yadav, B. S. 2013. Physicochemical, functional, thermal and pasting properties of starches isolated from pearl millet cultivars. *International Food Research Journal* 20: 1555–1565.

Kim, S. K., Choi, H. J., Kang, D. K. and Kim, H. Y. 2012. Starch properties of native proso millet (*Panicum miliaceum* L.). *Agronomy Research* 10: 311–318.

Kim, S. K., Sohn, E. Y. and Lee, I. J. 2009. Starch properties of native foxtail millet, *Setaria italica* Beauv. *Journal of Crop Science and Biotechnology* 12: 59–62.

Kim, W. W. and Yoo, B. 2009. Rheological behaviour of a corn starch dispersions: effects of concentration and temperature. *International Journal of Food Science & Technology* 44: 503–509.

Knutson, C. A., Khoo, U., Cluskey, J. E. and Inglett, G. E. 1982. Variation in enzyme digestibility and gelatinization behavior of corn starch granule fractions. *Cereal Chemistry* 59: 512–515.

Krueger, B. R., Knutson, C. A., Inglett, G. E. and Walker, C. E. 1987. A differential scanning calorimetry study on the effect of annealing on gelatinization behaviour of corn starch. *Journal of Food Science* 52: 715–718.

Kumari, S. K. and Thayumanavan, B. 1998. Characterization of starches of proso, foxtail, barnyard, kodo, and little millets. *Plant Foods for Human Nutrition* 53: 47–56.

Lawal, O. S., Lapasin, R., Bellich, B., Olayiwola, T. O., Cesàro, A., Yoshimura, M. and Nishinari, K. 2011. Rheology and functional properties of starches isolated from five improved rice varieties from West Africa. *Food Hydrocolloids* 25: 1785–1792.

Li, Q., Xie, Q., Yu, S. and Gao, Q. 2013. New approach to study starch gelatinization applying a combination of hot-stage light microscopy and differential scanning calorimetry. *Journal of Agricultural & Food Chemistry* 61: 1212–1218.

123

Li, K., Zhang, T., Sui, Z., Narayanamoorthy, S., Jin, C., Li, S. and Corke, H. 2019. Genetic variation in starch physicochemical properties of Chinese foxtail millet (*Setaria italica* Beauv.). *International Journal of Biological Macromolecules* 133: 337–345.

Lindeboom, N., Chang, P. R. and Tyler, R. T. 2004. Analytical, biochemical and physicochemical aspects of starch granule size, with emphasis on small granule starches: a review. *Starch/Stärke* 56: 89–99.

Liu, Q. 2005. Understanding starches and their roles in foods. In: Cui, S. W., editor. *Food carbohydrates: chemistry, physical properties, and applications*: Boca Raton: CRC Press, Taylor and Francis Group. pp. 327–329.

Liu, C., Liu, P. L., Yan, S. Q., Qing, Z. and Shen, Q. 2011. Relationship of physico-chemical, pasting properties of millet starches and the texture properties of cooked millet. *Journal of Texture Studies* 42: 247–253.

Lu, S., Chen, L. N. and Lii, C. Y. 1997. Correlations between the fine structure, physicochemical properties and retrogradation of amylopectins from Taiwan rice varieties. *Cereal Chemistry* 74: 34–39.

Lu, H., Zhang, J., Liu, K., Wu, N., Li, Y., et al. 2009. Earliest domestication of common millet (*Panicum miliaceum*) in East Asia extended to 10,000 years ago. *Proceedings of the National Academy of Sciences of the United States of America* 106: 7367–7372.

Madhusudhan, B. and Tharanathan, R. N. 1995. Legume and cereal starches—why differences in digestibility?—Part II. Isolation and characterization of starches from rice (*O. sativa*) and ragi (finger millet, *E. coracana*). *Carbohydrate Polymers* 28: 153–158.

Malleshi, N. G., Desikachar, H. S. R. and Tharanathan, R. N. 1986. Physico-chemical properties of native and malted finger millet, pearl millet and foxtail millet starches. *Starch/Stärke* 38: 202–205.

McCleary, B. V. and Monaghan, D. A. 2002. Measurement of resistant starch. *Journal of AOAC International* 85: 665–675.

Morales-Martínez, L. E., Bello-Pérez, L. A., Sánchez-Rivera, M. M., Ventura-Zapata, E. and Jiménez-Aparicio, A. R. 2014. Morphometric, physicochemical, thermal, and rheological properties of rice (*Oryza sativa* L.) Cultivars Indica× Japonica. *Food & Nutrition Sciences* 5: 271–279.

Naguleswaran, S., Vasanthan, T., Hoover, R. and Liu, Q. 2010. Structure and physicochemical properties of palmyrah (*Borassus flabellifer* L.) seed-shoot starch grown in Sri Lanka. *Food Chemistry* 118: 634–640.

Naidoo, K., Amonsou, E. O. and Oyeyinka, S. A. 2015. In vitro digestibility and some physicochemical properties of starch from wild and cultivated amadumbe corm. *Carbohydrate Polymers* 125: 9–15.

Naivikul, O. and Appolonia, B. L. D. 1979. Carbohydrates of legume flours compared with wheat flour. *Cereal Chemistry* 56: 24–28.

Noda, T., Takahata, Y., Sato Ikoma, H. and Mochida, H. 1996. Physicochemical properties of starches from purple and orange-fleshed sweet potato roots at two levels of fertilizer. *Starch* 48: 395–399.

Nwokocha, L. M. and Williams, P. A. 2011. Structure and properties of *Treculia africana*, (Decne) seed starch. *Carbohydrate Polymers* 84: 395–401.

Odeku, O. A. and Alabi, C. O. 2007. Evaluation of native and modified forms of *Pennisetum glaucum* (millet) starch as disintegrant in chloroquine tablet formulations. *Journal of Drug Delivery Science and Technology* 17: 155–158.

Perez, S., Baldwin, P. M. and Gallant, D. J. 2009. Structural features of starch granules I. In: BeMiller, J. and Whistler, R., editors. *Starch: chemistry and technology*. Amsterdam: Elsevier Inc.

Pérez, S. and Bertoft, E. 2010. The molecular structures of starch components and their contribution to the architecture of starch granules: a comprehensive review. *Starch/Stärke* 62: 389–420.

Punia, S., Siroha, A. K., Sandhu, K. S. and Kaur, M. 2019. Rheological and pasting behavior of OSA modified mungbean starches and its utilization in cake formulation as fat replacer. *International Journal of Biological Macromolecules* 128: 230–236.

Pycia, K., Juszczak, L., Gałkowska, D. and Witczak, M. 2012. Physicochemical properties of starches obtained from Polish potato cultivars. *Starch/Stärke* 64: 105–114.

Qi, Y., Wang, N., Yu, J., Wang, S., Wang, S. and Copeland, L. 2019. Insights into structure-function relationships of starch from foxtail millet cultivars grown in China. *International Journal of Biological Macromolecules* 155: 1176–1183.

Ratnayake, W. S., Hoover, R. and Warkentin, T. 2002. Pea starch: composition, structure and properties e a review. *Starch/Stärke* 54: 217–234.

Ring, S. G., Gee, J. M., Whittam, M., Orford, P. and Johnson, I. T. 1988. Resistant starch: its chemical form in foodstuffs and effect on digestibility in vitro. *Food Chemistry* 28: 97–109.

Rubens, P. and Heremans, K. 2000. Pressure-temperature gelatinization phase diagram of starch: an in situ Fourier transform infrared study. *Biopolymers* 54: 524–530.

Sandhu, K. S., Sharma, L., Kaur, M. and Kaur, R. 2020. Physical, structural and thermal properties of composite edible films prepared from pearl millet starch and carrageenan gum: process optimization using response surface methodology. *International Journal of Biological Macromolecules* 143: 704–713.

Sandhu, K. S. and Singh, N. 2005. Relationships between selected properties of starches from different corn lines. *International Journal of Food Properties* 8: 481–491.

Sandhu, K. S. and Siroha, A. K. 2017. Relationships between physicochemical, thermal, rheological and in vitro digestibility properties of starches from pearl millet cultivars. *LWT-Food Science & Technology* 83: 213–224.

Sandhu, K. S., Siroha, A. K., Punia, S. and Nehra, M. 2020a. Effect of heat moisture treatment on rheological and in vitro digestibility properties of pearl millet starches. *Carbohydrate Polymer Technologies and Application* 1: 100002.

Sang, Y., Bean, S., Seib, P. A., Pedersen, J. and Shi, Y. C. 2008. Structure and functional properties of sorghum starches differing in amylose content. *Journal of Agricultural & Food Chemistry* 56: 6680–6685.

Shaikh, M., Ali, T. M. and Hasnain, A. 2015. Post succinylation effects on morphological, functional and textural characteristics of acid-thinned pearl millet starches. *Journal of Cereal Science* 63: 57–63.

Shaikh, M., Ali, T. M. and Hasnain, A. 2017. Utilization of chemically modified pearl millet starches in preparation of custards with improved cold storage stability. *International Journal of Biological Macromolecules* 104: 360–366.

Sharma, M., Singh, A. K., Yadav, D. N., Arora, S. and Vishwakarma, R. K. 2016. Impact of octenyl succinylation on rheological, pasting, thermal and physicochemical properties of pearl millet (*Pennisetum typhoides*) starch. *LWT-Food Science & Technology* 73: 52–59.

Sharma, M., Yadav, D. N., Singh, A. K. and Tomar, S. K. 2015. Rheological and functional properties of heat moisture treated pearl millet starch. *Journal of Food Science & Technology* 52: 6502–6510.

Singh, M. and Adedeji, A. A. 2017. Characterization of hydrothermal and acid modified proso millet starch. *LWT-Food Science & Technology* 79: 21–26.

Singh, N., Kaur, L., Sandhu, K. S., Kaur, J. and Nishinari, K. 2006. Relationships between physical, morphological, thermal, rheological properties of rice starches. *Food Hydrocolloids* 20: 532–542.

Singh, J., Kaur, L. and Singh, N. 2004. Effect of acetylation on some properties of corn and potato starches. *Starch/Stärke* 56: 586–601.

Singh, N., Singh, J., Kaur, L., Sodhi, N. S. and Gill, B. S. 2003. Morphological, thermal and rheological properties of starches from different botanical sources. *Food Chemistry* 81: 219–231.

Singh, N., Singh, S. and Shevkani, K. 2011. Maize: composition, bioactive constituents, and unleavened bread. In: *Flour and breads and their fortification in health and disease prevention*. London: Academic Press: Elsevier. pp. 89–99.

Siroha, A. K., Punía, S., Sandhu, K. S. and Karwasra, B. L. 2020. Physicochemical, pasting, and rheological properties of pearl millet starches from different cultivars and their relations. *Acta Alimentaria* 49: 49–59.

Siroha, A. K. and Sandhu, K. S. 2018. Physicochemical, rheological, morphological, and in vitro digestibility properties of cross-linked starch from pearl millet cultivars. *International Journal of Food Properties* 21: 1371–1385.

Siroha, A. K., Sandhu, K. S., Kaur, M. and Kaur, V. 2019a. Physicochemical, rheological, morphological and in vitro digestibility properties of pearl millet starch modified at varying levels of acetylation. *International Journal of Biological Macromolecules* 131: 1077–1083.

Siroha, A. K., Sandhu, K. S. and Punia, S. 2019b. Impact of octenyl succinic anhydride on rheological properties of sorghum starch. *Quality Assurance and Safety of Crops & Foods* 11: 221–229.

Suh, D. S. and Jane, J. L. 2003. Comparison of starch pasting properties at various cooking conditions using the micro visco-amylo-graph and the rapid visco analyser. *Cereal Chemistry* 80: 745–749.

Suma P., F. and Urooj, A. 2015. Isolation and characterization of starch from pearl millet (*Pennisetum typhoidium*) flours. *International Journal of Food Properties* 18: 2675–2687.

Sun, Q., Gong, M., Li, Y. and Xiong, L. 2014. Effect of dry heat treatment on the physicochemical properties and structure of proso millet flour and starch. *Carbohydrate Polymers* 110: 128–134.

Suriya, M., Baranwal, G., Bashir, M., Reddy, C. K. and Haripriya, S. 2016. Influence of blanching and drying methods on molecular structure and functional properties of elephant foot yam (*Amorphophallus paeoniifolius*) flour. *LWT-Food Science & Technology* 68: 235–243.

Tester, R. F. and Morrison, W. R. 1990. Swelling and gelatinization of cereal starches. I. Effects of amylopectin, amylose, and lipids. *Cereal Chemistry* 67: 551–557.

Toker, O. S., Dogan, M., Canıyılmaz, E., Ersoz, N. B. and Kaya, Y. 2013. The effects of different gums and their interactions on the rheological properties of a dairy dessert: a mixture design approach. *Food & Bioprocess Technology* 6(4): 896–908.

Wang, S. and Copeland, L. 2015. Effect of acid hydrolysis on starch structure and functionality: a review. *Critical Reviews in Food Science & Nutrition* 55(8): 1081–1097.

Wang, S., Li, C., Copeland, L., Niu, Q. and Wang, S. 2015. Starch retrogradation: a comprehensive review. *Comprehensive Reviews in Food Science & Food Safety* 14: 568–585.

Wani, A. A., Singh, P., Shah, M. A., Schweiggert-Weisz, U., Gul, K. and Wani, I. A. 2012. Rice starch diversity: effects on structural, morphological, thermal, and physicochemical properties–a review. *Comprehensive Reviews in Food Science & Food Safety* 11(5): 417–436.

Wani, I. A., Sogi, D. S., Wani, A. A., Gill, B. S. and Shivhare, U. S. 2010. Physicochemical properties of starches from Indian kidney bean (*Phaseolus vulgaris*) cultivars. *International Journal of Food Science & Technology* 45: 2176–2185.

Wani, I. A., Sogi, D. S., Hamdani, A. M., Gani, A., Bhat, N. A. and Shah, A. 2016. Isolation, composition, and physicochemical properties of starch from legumes: A review. *Starch/Stärke* 68: 834–845.

Wankhede, D. B., Rathi, S. S., Gunjal, B. B., Patil, H. B., Walde, S. G., Rodge, A. B. and Sawate, A. R. 1990. Studies on isolation and characterization of starch from pearl millet (*Pennisetum americanum* (L.) Leeke) grains. *Carbohydrate Polymers* 13: 17–28.

Wankhede, D. B., Shehnaz, A. and Raghavendra Rao, M. R. 1979. Preparation and physicochemical properties of starches and their fractions from finger millet (*Eleusine coracana*) and foxtail millet (*Setaria italica*). *Starch/Stärke* 31(5): 153–159.

Wu, Y., Lin, Q., Cui, T. and Xiao, H. 2014. Structural and physical properties of starches isolated from six varieties of millet grown in China. *International Journal of Food Properties* 17: 2344–2360.

Yanez, G. A., Walker, C. E. and Nelson, L. A. 1991. Some chemical and physical properties of proso millet (*Panicum milliaceum*) starch. *Journal of Cereal Science* 13(3): 299–305.

127

Zhao, K. and Gu, G. 2007. Research advances in starch digestibility analysis. *Food Science* 28(482): 586–590.

Zhong, F., Li, Y., Ibáñez, A. M., Oh, M. H., McKenzie, K. S. and Shoemaker, C. 2009. The effect of rice variety and starch isolation method on the pasting and rheological properties of rice starch pastes. *Food Hydrocolloids* 23(2): 406–414.

Zhou, X., Baik, B. K., Wang, R. and Lim, S. T. 2010. Retrogradation of waxy and normal corn starch gels by temperature cycling. *Journal of Cereal Science* 51(1): 57–65.

Zhu, F. 2016. Buckwheat starch: structures, properties, and applications. *Trends in Food Science & Technology* 49: 121–135.

Zobel, H. F. 1988a. Starch crystal transformations and their industrial importance. *Starch/Stärke* 40: 1–7.

Zobel, H. F. 1988b. Molecules to granules-a comprehensive starch review. *Starch* 40: 44–50.

Zobel, H. F., Young, S. N. and Rocca, L. A. 1988. Starch gelatinization: an X-ray diffraction study. *Cereal Chemistry* 65(6): 443–446.

6

Impact of Modification on Starch Properties

6.1 INTRODUCTION

Starch, a natural carbohydrate polymer, acts as the most useful ingredient in food industries. Regardless of the key functional roles of starch, native starch sometimes affects several quality characteristics of food products and causes instability of the paste under shear, acid, or freezing conditions and poor paste clarity (Shaikh et al., 2015). Therefore, the applications of starch in various industries like food, paper, and textile can be increased by adopting various techniques of modification. Modification is defined as the process that brings changes in starch structure by various factors, i.e., environmental, operational, processing, etc. The structure and functionality of starch molecules are affected by modifications either positively or negatively. In order to promote and enhance specific functional properties, starches are frequently modified by physical, chemical, and enzymatic processes (Zia-ud-Din et al., 2017). Physical methods use thermal and nonthermal techniques, chemical modifications introduce functional groups into the starch molecule using derivatization reactions (e.g., etherification, esterification, crosslinking) or involve breakdown reactions (e.g., hydrolysis and oxidation) (Singh et al., 2007), and enzymatic modification involves enzymes. The rate and efficiency of these methods are dependent on the reagent type, primary biomaterial source, and the corresponding physical attributes (size and structure) of starch granules

(Huber & BeMiller, 2001). The utilization of modified starches in food and pharmaceutical industries has continued to increase thanks to their broadened and improved physicomechanical and functional properties compared to native starches (Kittipongpatana & Kittipongpatana, 2013).

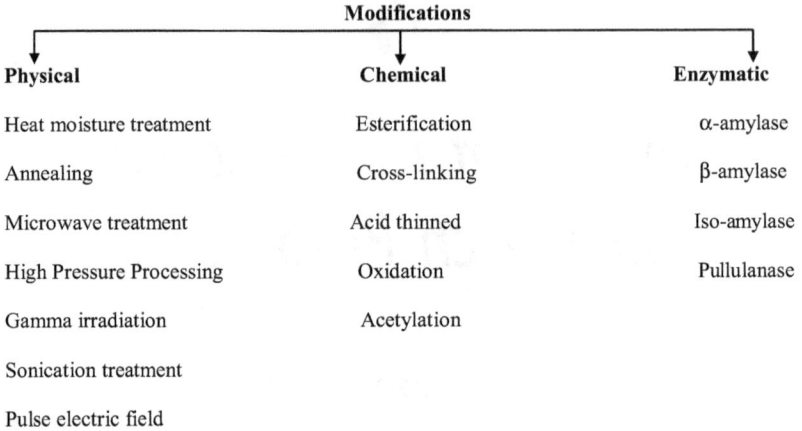

Modifications		
Physical	**Chemical**	**Enzymatic**
Heat moisture treatment	Esterification	α-amylase
Annealing	Cross-linking	β-amylase
Microwave treatment	Acid thinned	Iso-amylase
High Pressure Processing	Oxidation	Pullulanase
Gamma irradiation	Acetylation	
Sonication treatment		
Pulse electric field		

6.2 PHYSICAL MODIFICATION

Food industries have shown their interest in producing more natural food components; so, there is an increasing need to improve the properties of native starches without using chemical modifications (Ortega-Ojeda & Eliasson, 2001). Among various physical modification methods, high hydrostatic pressure (HHP), a non-thermal processing method, is considered to be a suitable technique for the production of minimally processed foods (Colussi et al., 2020). Different researchers reported physical modifications of millet starch, which are shown in Table 6.1. These modifications are taken into consideration as these are simple, cost-effective, eco-friendly, and safe methods and because chemicals or biological agents, which are harmful for human consumption, are not required (Punia, 2020).

6.2.1 Heat Moisture Treatment

Heat moisture treatment (HMT) is a physical method that is carried out under restricted moisture content (10%–30%), at higher temperatures (90°C–120°C), and for a period ranging from 15 min to 16 h (Maache-Rezzoug

Table 6.1 Physical Modification of Millet Starch

Modification	Major Results and Methodology	References
HMT	Decrease in pasting properties and increase in gelatinization temperature were observed. Less than 1% starch granule affected by HMT (finger millet starch (20%–30% moisture content) heated for 16 h at 100°C).	Adebowale et al. (2005)
HMT	Swelling power, solubility FV, and SV increased; PV, TV, BV, PT decreased (pearl millet starch (25%–28% moisture content) heated for 3 h at 110°C).	Balasubramanian et al. (2014)
HMT	Swelling power, solubility, PV, and BD increased while gelatinization of pearl millet starch with 20%, 25%, and 30% moisture content heated at 110±2°C for 8 h.	Sharma et al. (2015)
HMT	Relative crystallinity, pasting properties, and thermal enthalpy of modified starches decreased (proso millet starch (20 g, dry basis) and water (80 ml) were mixed well. The HMT sample was heated for 30 min at 100°C). Physicochemical, pasting, and rheological properties reduced after modification. After modification, RS and SDS contents increased and RDS content decreased as compared to their native counterpart starches. Pearl millet starches were modified using HMT (moisture content 25%, 110°C at 3 h).	Zheng et al. (2020), Sandhu et al. (2020)
Annealing	Decrease in pasting properties and increase in gelatinization temperature were observed. Less than 1% starch granule affected by annealing treatment (heating finger millet starch at 50°C for 48 h).	Adebowale et al. (2005)
Annealing	Resistant starch content, shear stability, and amylose content increased as compared to native starch. SEM shows granules fusion and development of pores after modification (heating finger millet starch at 50°C for 24 h).	Babu et al. (2019)

(Continued)

Table 6.1 (*Continued*) Physical Modification of Millet Starch

Modification	Major Results and Methodology	References
Microwave	Microwave treatment caused a decrease of the≈peak viscosity, swelling power, ΔH, and relative crystalline and an increase of the transparency, T_o, T_C, and ΔT (millet starch with moisture content (30%–50%), microwave irradiation was carried out for 60 s and the microwave power was 700 W).	Li et al. (2019b)
Microwave	Swelling power and relative crystallinity decreased, while transparency increased. Microwave also destroyed the original appearance of the starch granules and formed smaller and lamellar gel blocks. (microwave treatment: millet starch slurry (10%, 15%, 20%, 25%, and 30%) was heated by microwave for 30, 60, 90, and 120 s.)	Li et al. (2019b)
Microwave	Swelling power, solubility, pasting properties (peak, trough, break down, and final and pasting temperature) and molecular weight decreased (proso millet starch is heated at a microwave power of 500 W for 10 min).	Zheng et al. (2020)
HHP	Gelatinization temperature increases, while the reverse is observed for ΔH when starch is treated at 150–450 MPa. Pasting properties also decreased (proso millet starch (30% w/v) treated at pressure levels of 150, 300, 450 and 600 MPa for 15 min using a high-pressure machine).	Li et al. (2018)
Sonication	Swelling power, transmittance, and thermal measurement increased. No significant effect of sonication is observed on pasting properties (ultrasound treatment: millet starch slurry was placed in the ultrasonic sink for 15, 30, 45 and 60 min).	Li et al. (2019b)

et al., 2008). Figure 6.1 describes the method of HMT at different temperatures. HMT raises the gelatinization temperature and expands its span while decreasing the swelling factor and amylose leaching in all kinds of starches. However, the effects of HMT on the transition of crystalline

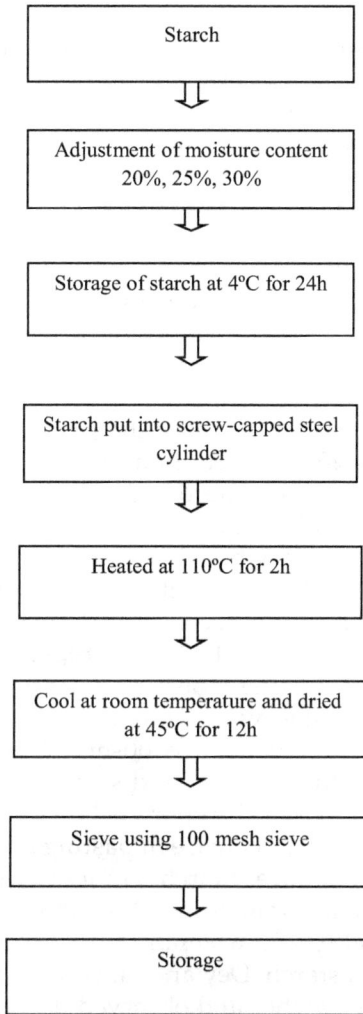

Figure 6.1 Method for heat moisture treatment (Liu et al., 2019).

type, crystallinity interruption, and change of enzyme susceptibility differ depending on the starch source (Gunaratne & Hoover, 2002). The amount moisture content during HMT treatment is an important factor to alter the various properties of starches. Water in starch samples acts as a plasticizer,

133

rendering starch polymeric chains more flexible and thus facilitating the rearrangement of amylose/amylopectin unit chains (Olayinka et al., 2008; Watcharatewinkul et al., 2009). HMT has been shown to increase thermal stability and shear resistance of starches (Hoover, 2010) and, therefore, can be used as an alternative to chemically modified starches in retort foods, confections, salad dressings, and batter products (Hoover & Vasanthan, 1994).

Depending on the origin of starch, the polymorph of HMT starch may differ from that of the native one, which is the A type crystalline structure, remains unaltered but the B-type polymorphis transformed to C- or A-type structure. In addition, an unaltered or reduced crystallinity can be observed for HMT starch of different origins (Lin et al., 2019). HMT induced structural rearrangement by disrupting the crystalline structure and dissociating the double helical structure in the amorphous region without disturbing the granular structure of starch (Gunaratne & Hoover, 2002). Disruption of starch crystallites and rearrangement of starch molecules occurred simultaneously during HMT. Therefore, appropriate moisture content and temperature could lead to higher relative crystallinity, whereas relatively high moisture content and temperature could enhance the effectiveness of HMT at disrupting starch crystallites and result in decreased relative crystallinity (Li et al., 2011; Wang et al., 2017). Sharma et al. (2015) reported HMT modification of pearl millet starch at varying moisture contents (20%, 25%, and 30%). It was observed that G' value, yield, and flow point were decreased after the modification and the reverse was observed for gelatinization temperature. HMT also induced cavity and some dents on starch granule surface. Adebowale et al. (2005) reported HMT modification of finger millet starch and observed a decrease in pasting properties and increase in gelatinization temperature. Starch granules were slightly affected by HMT treatment. Balasubramanian et al. (2014) observed increase in swelling power, solubility, and water binding capacity after HMT modification for pearl millet starch. Dey and Sit (2017) reported HMT modification of foxtail millet starches and observed decreased swelling power and pasting properties. Zheng et al. (2020) evaluated the HMT treatment of proso millet starch and observed that relative crystallinity of HMT starch decreased as compared to native counterpart starch. Pasting properties were decreased after HMT.

HMT promoted the increase of slowly digestible starch (SDS) and resistant starch (RS) content and was dependent on the degree of moisture content and treatment time during HMT (Chen et al., 2017). HMT could transform part of the rapidly digested starch (RDS) into SDS and/or RS (Liu et al., 2019). SDS and RS presented a relatively low glycemic response, which could contribute to suppressing metabolic diseases (Wang et al., 2017). Interestingly, HMT can transform rapidly digestible and resistant starch fractions into the slowly digestible forms (Zeng et al., 2015; Tan et al., 2017).

6.2.2 Annealing

Annealing occurs when starch granules are heated (below gelatinization temperature but above glass transition temperature) in an excess (> 60%, w/w) or intermediate (40%–55%, w/w) amount of water content (Jayakody & Hoover, 2008; Ashogbon & Akintayo, 2014). Annealing has been reported to cause a reorganization of starch molecules, leading to amylopectin double helices to acquire one more organized configuration (Krueger et al., 1987; Hoover & Vasanthan, 1994). Annealing also leads to elevation and sharpening of the gelatinization range and causes little solubilization of the α-glucan (Tester & Debon, 2000). Annealing makes significant changes in physicochemical and functional properties of starch. It includes decrease in swelling, solubility, and gelatinization parameters and increase in paste stability, crystallinity, and enzyme susceptibility (Wang et al., 2013). Annealing can be carried out in different steps, the most significant being the single-step annealing (Vermeylen et al., 2006). The utilization of the double-stage annealing (Genkina et al., 2007; Shi, 2008) and multistep annealing (Nakazawa & Wang, 2003, 2004) has been conducted to a restricted extent. The annealing results in a more perfectly ordered structure and in an increase in the granule stability by improving the arrangement of double helices as well as the perfection of starch crystallites (Jayakody & Hoover, 2008). The effects of annealing are maximized at temperatures immediately below the onset gelatinization temperature (Tester & Debon, 2000). Adebowale et al. (2005) reported the annealing modification of finger millet starch. Pasting properties of annealed starch decreased as compared to native starch. Annealing did not alter the shape or the surface characteristics of the starch granules appreciably. Babu et al. (2019) evaluated annealing of foxtail millet starch and observed that after annealing process relative crystallinity of starch increases as compared to native starch. After annealing process, RDS content decreased, whereas increase in SDS and RS contents was observed.

6.2.3 Microwave Treatment

Microwave irradiation is one of the physical methods of starch modification of notable interest in research. Among the three (physical, chemical, and enzymatic) main treatments, physical modification of starch has become of immense importance as it is a green technology with noninvolvement of chemicals (Kumar et al., 2020). The method of microwave treatment for starches is shown in Figure 6.2.

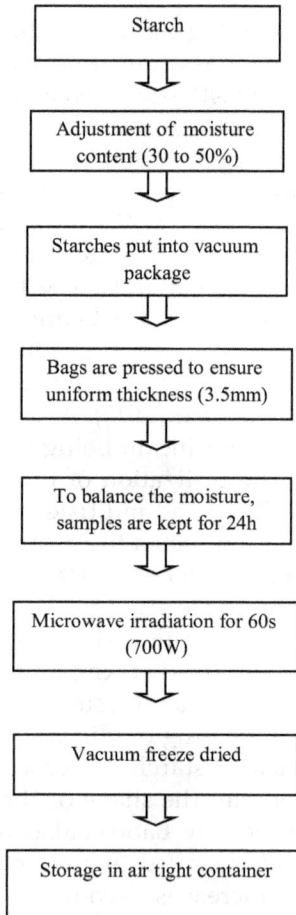

Figure 6.2   Method for preparation of microwave treatment of starch (Li et al., 2019a).

Microwave irradiation, *i.e.*, treatment with electromagnetic waves with frequencies ranging from 300 MHz to 300 GHz (Salazar-González et al., 2012), is commonly adopted in food processing for heating, baking, enzyme deactivation, modification, etc. (Fan et al., 2013). Most of the studies were found to involve high-power microwave modification of starch: 1000 W (Shen et al., 2017), 450 W (Arocas et al., 2011), 1100 W (Colussi et al., 2017), sequential heating at 1000 W, 350 W, and finally at 650 W (Ma et al., 2015; Fan et al., 2017) vacuum microwave treatment at 460, 500, and 750 W at 3800 Pa (Mollekopf et al., 2011), 450 W (Nadir et al., 2015), 800 W-5 min (Xia et al., 2018), 2450 MHz, and 750 W (Xie et al., 2013). Microwaves are nonionizing electromagnetic radiations capable of inducing changes in the properties of materials due to rapid alternations of the electromagnetic field at high frequency (Deka & Sit, 2016). Different from the conduction heating modes, microwave can make the materials achieve a higher temperature in a shorter time by the rapid vibration of polar molecules in the microwave fields (Acierno et al., 2004). Processing with microwave is more efficient than the traditional heating processes due to its shorter processing time and ensures homogenous operation in the whole volume of substance (Anderson & Guraya, 2006). Microwave treatment affects starch through dielectric heating and electromagnetic polarization effects (Bilbao-Sáinz et al., 2007). Dielectric properties of starches reflect their response to microwaves and their ability to convert microwave energy into heat energy, thereby determining the penetration depth of microwaves in them. Common measurement methods for the dielectric properties of foodstuffs include open coaxial probe technology (Fan et al., 2017; Auksornsri et al., 2018). Several factors such as moisture content of starch, microwave heating time, microwave heating power, and the botanical source may significantly influence the physicochemical properties of starch (Oyeyinka et al., 2019). Braşoveanu and Nemţanu (2014) stated that starch is affected by microwave and starch factor. Microwave factors include the frequency, power, radiation time, and geometry of heating system. Starch factors include the starch types, density, and dielectric properties. The effects of microwave interaction with matter are based on the microwave energy which is delivered directly to the material through molecular interaction with the electromagnetic field (Braşoveanu & Nemţanu, 2014). Li et al. (2019a) evaluated the microwave treatment of millets at different moisture content. Microwave treatment caused a decrease of the peak viscosity, swelling power, ΔH, and relative crystalline and an increase of the transparency,

T_O, T_C, and ΔT. Cracks and center cavities were observed on the granule surface, and the original appearance of starch disappeared gradually as the moisture content increases. Luo et al. (2006) also stated that microwave irradiation causes a shift in the gelatinization range to higher temperatures and a drop in viscosity and crystallinity, which may be due to the molecular rearrangements restricted to sections of the starch molecules. Zheng et al. (2020) studied microwave treatment of proso millet starch and observed that swelling power, solubility, pasting properties (peak, trough, breakdown, and final and pasting temperature) and molecular weight are decreased as compared to native starch. Amylose content and RS content increased after microwave treatment. Microwaves can breakdown starch to smaller fragments by damaging the glycosidic bonds (Shah et al., 2016). Li et al. (2019b) reported microwave treatment of millet starch. It is observed that after swelling treatment, relative crystallinity is decreased as compared to native starch while *in vitro* digestibility is increased. Gelatinization temperature is increased while decrease in enthalpy of gelatinization is observed. SEM shows that granules are destroyed after treatment.

6.2.4 High-Pressure Processing

Consumers worldwide are interested in non-thermal food processing methods in order to preserve nutritive compounds and to reduce the use of chemical food additives (Tokuşoğlu & Swanson, 2014). High-pressure processing (HPP) is a nonthermal processing technology that has been employed successfully for the gelatinization or modification of starch (Ahmed et al., 2017). HPP is a green and environmental-friendly technology and can alter noncovalent chemical linkages with minimal effects on covalent linkages; thus, it can be used to modify starch in order to have tailor-made desired properties, since the native starch does not have the suitable properties for processing (Castro et al., 2020). In HHP processing, foods are subjected to a high pressure ranging from 200 to 600 MPa for a given time, with water as the pressure transmitting medium (San Martin et al., 2002). Compared with traditional thermal processing technology, HPP is performed at room temperature, reducing energy consumption associated with heating and subsequent cooling. In HPP treatment, the food is in packaged form and does not directly contact the processing devices, preventing the secondary contamination of food after pasteurization (Huang et al., 2017). The HHP treatment has been applied in many processed foods, such as gelatinization and physical modification of starch from various botanical sources (Bruschi et al., 2017). The extent of pressure-assisted starch gelatinization/gelation depends upon many factors including pressure intensity, holding time, starch concentration, temperature, and type of starches or preparation techniques (Ahmed et al., 2016). HPP treatment provokes swelling of starch but maintains granule integrity; as a consequence, HPP-treated starches modify their microstructure and rheological properties in a different way than thermally treated ones (Gomes et al., 1998). HPP treatment has been shown to influence mostly the noncovalent bonds in starches, so this treatment can cause major structural, textural, sensory, and nutritional damages (Balny et al., 2002). Starch granules could be gelatinized completely at room temperature by HPP, and the impact of the process in starch gelatinization depends markedly on starch type, treatment pressure, temperature, time, and water content (Bauer & Knorr, 2005; Li et al., 2012). However, the mechanisms of the high-pressure induced starch gelatinization differed from the gelatinization that was achieved by heating starch granules in the presence of water. During the HHP treatment of the starch water suspension, the water molecules were penetrating from amorphous into crystalline domains by high pressure, resulting in an irreversible granule swelling, which is defined as pressure-induced gelatinization (Błaszczak et al., 2005). The starch gelatinization degrees and rates under a given HHP treatment condition seemed to be mainly influenced by the inherent structures and compositions of starch granules (Kim et al., 2012). It is reported that HHP treatment has a range of effects on these starches, including improving the thermal stability (Douzals et al., 1998), decreasing the swelling index, lowering the gelatinization temperature and gelatinization enthalpy (Kawai et al., 2012), weakening starch gels (Stolt et al., 2000), and altering the crystalline structure and morphology (Błaszczak et al., 2005). Li et al. (2018) reported ultra-high-pressure treatment for proso millet starch. It is observed that when starch is treated with 600 MPa, XRD pattern of starch is changed from A-type to B-type. Gelatinization temperature of modified starches increase, while the reverse is observed for gelatinization enthalpy when starch is treated at 150–450 MPa. Pasting property, i.e. trough and final viscosity, pasting temperature, and peak time are decreased; however, peak and breakdown viscosity are decreased with the increase in pressure level up to 600 MPa.

139

6.2.5 Sonication Treatment

The ultrasound field was established in the 1880s and found the first commercial application in 1917 with the echo-sounding technique (Lorimer & Mason, 1987). Ultrasonication showed beneficial effects in food processing and preservation, including higher product yields, shorter processing times, reduced operating and maintenance costs, improved quality attributes, reduction of pathogens, and so on (Patist & Bates, 2008). Ultrasonic treatment is a physical method of modifying starch that offers advantages in terms of the reduced use of chemicals and less time consumed by processing and, as such, is more environmentally acceptable (Monroy et al., 2018). In order to use ultrasound in starch modification, either the native starch solution or starch after gelatinization can be subjected to the ultrasound (Zuo et al., 2009). The process of sonication treatment of starch is shown in Figure 6.3. Ultrasound means the mechanical waves with a frequency exceeding 16 kHz, which can be transformed to "micro-jets", "high shearing force and turbulence," and "high pressure" through the cavitation effect, causing the molecular degradation and inherent structural transformations of starch granules (Lu et al., 2018; Yang et al., 2019a; Falsafi et al., 2019). This, in turn, may cause shear forces that have no significant influence on small molecules but are capable of breaking the chains of polymers, provided the chains are longer than a certain limiting value. Gedanken (2004) also stated that the polymer chains near the collapsing microbubbles are caught in a high-gradient shear field, which leads to the breakage of macromolecular C–C bonds and formation of long chain radicals. Tomasik and Zaranyika (1995) stated that the mechanism involves the formation of bubbles in the dispersion medium, which bombard starch granules before they collapse. This is the mechanochemical action of ultrasound on polymers (Czechowska-Biskup et al., 2005). Depending on the botanical origin, the physical state of the starch, the concentration of the slurry, and the characteristics of the sonication process (i.e., power, frequency, time, and temperature), sonicated starch undergoes a series of changes, including the breakage of the granular structure, a decrease in the viscosity, and an increase in the solubility and transparency (Zhu et al., 2012; Amini et al., 2015; Majzoobi et al., 2015; Zhu, 2015). By virtue of its cavitation and mechanical oscillation, ultrasound at an appropriate intensity and frequency can produce various modified starches with different functions (Seguchi & Hignsa, 1994). According to many

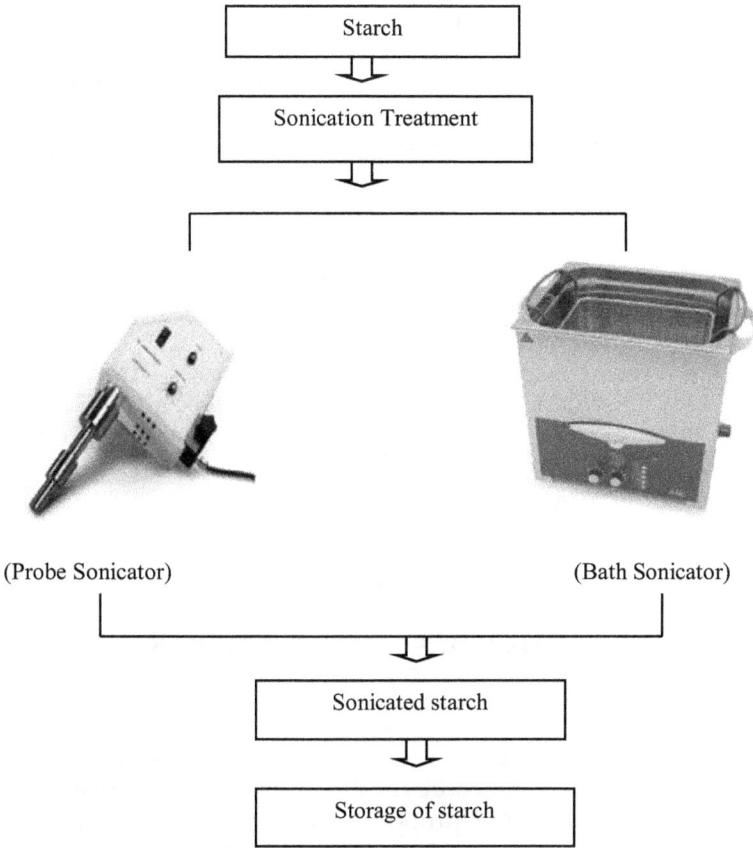

Figure 6.3   Process of sonication treatment of starch.

reports, ultrasounds cause physical degradation of granules with visible fissures and pores on the surface, but no alterations of granule shape and size have been observed (Luo et al., 2008; Zheng et al., 2013; Sujka & Jamroz, 2013; Amini et al., 2015). Li et al. (2019b) evaluated modification of millet starch by ultrasonication treatment. It is observed that ultra-sonication treatment led to increase in swelling power, transmittance, and thermal measurement. In the pasting properties, not much effect of sonication was observed, while relative crystallinity is decreased after treatment. Sujka and Jamroz (2013) reported that ultrasound treatment reduces the pasting viscosity of starch and increases the content of amylose and RS; besides, pores and cracks also appear on the surface of starches after ultrasound treatment. Ultrasonication treatment did not completely damage the original appearance of the starch granules and only brought out limited changes, such as the formation of cracks and pores. The reason was due to the high shear force, mechanical force, and pressure gradients generated by the fast formation and collapse of cavitation bubbles during ultrasonication (Yang et al., 2019). Li et al. (2009) reported the increase in enthalpy of gelatinization and *in vitro* enzymatic digestibility of starch after the ultrasonication treatment. The increased treatment of starch dispersions by ultrasound energy produced a large amount of short-chained molecules, which were more amenable for enzymatic degradation (Flores-Silva et al., 2017). Jambrak et al. (2010) studied ultrasound treatment of corn starch, finding that ultrasound waves distorted the crystalline region in starch granules and hardly changed the gelatinization properties.

6.3 CHEMICAL MODIFICATION

Chemical modification is a classical way to effectively improve the functionalities of starch. Native starches contain free hydroxyl groups in the 2, 3, and 6 carbons of the glucose molecule, making them highly reactive. This allows them to be modified by different chemical treatments, thus regulating their properties (Bao et al., 2003). The chemical modification of starch can be achieved by a variety of different chemical reactions such as acid hydrolysis, oxidation, etherification, esterification, and cross-linking (Jayakody & Hoover, 2002). Chemical modification of starches is an approach to markedly alter starch functionality and enhance its versatility by introduction of new functional groups, which in turn changes the physicochemical properties of starches (Gao et al., 2014). The chemical and functional properties achieved, upon modifying starch by chemical substitution, depend on starch source, reaction conditions (reactant concentration, reaction time, pH, and the presence of catalyst), type of substituent, extent of substitution (DS), and the distribution of the substituent in the starch molecule (Kavitha & BeMiller, 1998; Hirsch & Kokini, 2002). Studies on chemical modification of millet starches are listed in Table 6.2.

Table 6.2 Chemical Modification of Millet Starch

Modification	Major Results and Methodology	References
Cross-linking	Swelling power, solubility, pasting, and rheological properties decreased. Slight fragmentation in starch granule is observed after modification of pearl millet starch (0.5% EPI suspended in 0.5% O solution, agitating for 5 h and neutralizing, washing and centrifugation).	Siroha and Sandhu (2018)
Succinylation	Swelling power, solubility, PV, BD, and SB increased and PT decreased. Gelatinization temperatures (T_o, T_p, and T_c) decreased and ΔHgel value increased as compared to native pearl millet starch (succinic anhydride (4%) pH 9.0–9.5, time 2 h).	Shaikh et al. (2015)
Succinylation	Swelling power, PV, cool paste viscosity, hot paste viscosity, and BD increased, while SB and PT decreased. Gelatinization parameter (T_o, T_p, T_c, and ΔHgel) decreased. Yield stress and consistency index values increased as compared to native pearl millet starch (octenyl succinic anhydride (3%) pH 8, time 2, 3, 4, and 5 h).	Sharma et al. (2016)
Succinylation	Swelling power, pasting properties, and rheological properties increased after OS modification of foxtail millet starch. RDS content increased, while the reverse was observed for RS and SDS contents. Succinic anhydride treatment created cracks on the surface of the starch granule (starch slurry 40%, pH 9.5, 4% (w/w) succinic anhydride, time 2 h).	Babu and Mohan (2019)
Acetylation	Decrease in swelling power and solubility is observed. After modification of pearl millet starch, decreases in gelatinization temperature and enthalpy of gelatinization are observed (acetic anhydride 6%; pH 8.0–8.5; time 15 min).	Shaikh et al. (2017)
Acetylation	Gelatinization temperature and enthalpy of gelatinization decreased after the acetylation process. Acetylation modification shows less smooth surface with increased indentations on the surface.	Shaikh et al. (2019)

(Continued)

Table 6.2 (*Continued*)   Chemical Modification of Millet Starch

Modification	Major Results and Methodology	References
Acetylation	Swelling power, pasting properties, and rheological properties increased as compared to native counterpart pearl millet starch. After modification, a decrease in RDS and SDS content is observed, while the reverse is observed for RS content (acetic anhydride concentration (1.25%, 2.5%, 3.75% & 5.0%); time 10 min; pH 8.0).	Siroha et al. (2019)
Oxidation	Swelling power decreases, while increase in solubility is observed. PV, hot paste viscosity, cold paste viscosity, and SV increase, while BV and PT decrease as compared to native starch (pearl millet). Gelatinization temperature also decreases (20% starch slurry (total starch 150 g); pH 9.0–9.5; sodium hypochlorite 5 g, time, 45 min).	Shaikh et al. (2017)
Acid thinning	Swelling power, PV, BV, SV, and PT decreased and gelatinization parameters (T_o, T_p, T_c, and ΔHgel) increased when pearl millet starch is modified with 1.0 M HCl concentration (0.1 and 1.0 M HCl at 50°C for 2 h).	Shaikh et al. (2015)
Acid thinning	Swelling power, pasting parameters, and gelatinization temperature (T_o and T_p) decreased after the modification of proso millet starch (HCl 0.14 mol/L; 8 h; temperature 50°C).	Singh and Adedeji (2017)

6.3.1 Cross-Linking

Cross-linking means that molecules of polymer are interconnected by some sort of bonding that may be covalent, ionic, or hydrogen bonding, resulting from intermolecular forces (Mehboob et al., 2020). Common cross-linking agents currently used are sodium trimetaphosphate (STMP), sodium tripolyphosphate (STPP), epichlorohydrin (EPI), phosphoryl chloride ($POCl_3$), a mixture of adipic acid and acetic anhydride, and vinyl chloride (Woo & Seib, 1997). The method for preparation of cross-linked starch using EPI is described in Figure 6.4. EPI has a toxic nature; so, STMP has been proposed as a nontoxic and effective cross-linking agent for starch (Li et al., 2009). Cross-linking reinforces the hydrogen bonds in the granule

144

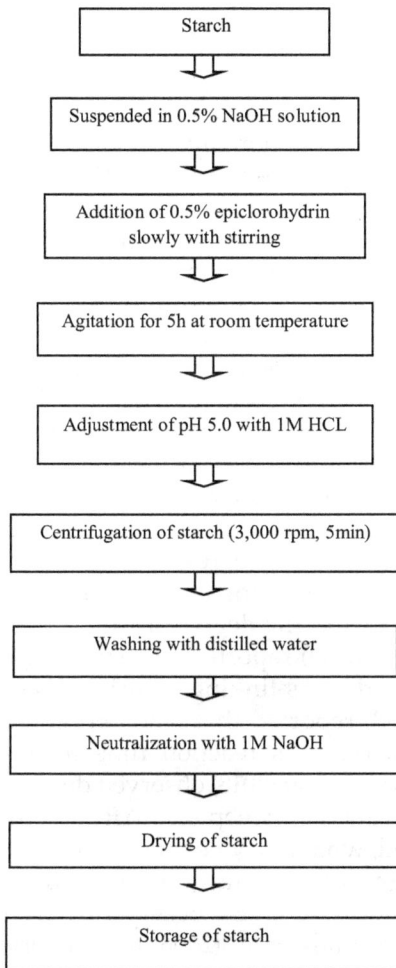

Figure 6.4 Method for preparation of cross-linked starch (Wurtzburg, 1960).

with chemical bonds that act as a bridge between the starch molecules (Jyothi et al., 2006). Cross-linking of starch was reported to be affected by various factors including starch source, cross-linking reagent concentration and composition, the extent of substitution, pH, reaction time, and temperature (Lim & Seib, 1993; Chung et al., 2004). Cross-linking alters not only the physical properties but also the thermal transition characteristics

145

of starch, and the effect of cross-linking depends on the botanical source of starch and the cross-linking agent (Solarek, 1986). In the food industry, starch modification with cross-linking is commonly used to provide stabilized granular structure and restricted swelling (Ratnayake & Jackson, 2008) as well as nutritionally beneficial effects (Wurzburg, 1986; Woo, 1999). Cross-linking via bi- or polyfunctional reagents reinforces the already present hydrogen bonds in granules with new covalent bonds. As a result, cross-linked starch is more resistant to acid, heat, and shearing than its native counterpart (Kaur et al., 2006). The relative effects of various reaction parameters in determining the extent of starch modification in terms of total level of functional groups incorporated, i.e. DS and/or difference in swelling/gelatinization/pasting properties between the cross-linked and native form of starches, have been extensively investigated (Hong et al., 2015). Siroha and Sandhu (2018) evaluated the cross-linking modification of pearl millet starch and observed degrees of cross-linking ranging from 40.8% to 89.7%. Swelling power, solubility, and pasting properties (pasting temperature) decreased as compared to the native counterpart starch. Cross-linking strengthens the bonding between the starch chains, causing an increase in resistance of the granules to swelling with increasing degree of cross-linking (Singh & Adedeji, 2007). Kurakake et al. (2009) reported that swelling of starch granules is depressed by enhanced cross-linking, which causes reduction in viscosity. Deetae et al. (2008) reported that reaction time of cross-linking also influenced peak viscosity. As reaction time increased, peak viscosity decreased. Siroha and Sandhu (2018) observed decrease in G' value during heating as well as frequency sweep test. After modification, decrease in peak G' was observed, which may be due to the lower degree of swelling and thus less inter-granular interaction. Yield stress, consistency index, and flow behavior index were also observed to have values lesser than those of native starch. Kaur et al. (2006) also reported that cross-linked potato starches show lower G' value as compared to their native starch. Kim and Yoo (2010) reported decreased values of consistency index and yield stress after the cross-linking modifications.

Morphological properties of starches are affected by the cross-linking modification of starches. Siroha and Sandhu (2018) reported that after cross-linking, morphological properties of starches changed as compared to native starches. Cross-linked starch granules, slightly rough surface, and the formation of grooves were observed, indicating slight fragmentation. Koo et al. (2010) observed that cross-linking modification caused slight changes in the structure of starch granules compared with native counterpart for corn starch. Carmona-Garcia et al. (2009) also found that cross-linked banana starch with STMP/STPP showed black zones on the surface along the longitudinal axis. Singh and Adedeji (2007) suggested that the black zones seemed to indicate a slight fragmentation and the formation of a deep groove in the starch granules.

Xie and Liu (2004) indicated that cross-linked starches cannot be completely digested since the substituents hinder enzymatic attack. For a healthy diet, it is recommended to decrease fat intake and increase dietary fiber consumption (Wolf et al., 1999). The health-beneficial role of RS is due to its fermenting ability by the colonic microflora, promoting production of short-chain fatty acids such as acetate, propionate, and butyrate (Aparicio-Saguilán et al., 2008). Siroha and Sandhu (2018) reported increase in the RS content after cross-linking modification. Koo et al. (2010) stated that the swelling of starch granule is likely to enhance the access of digestive enzymes to its inner structure; it appeared that the cross-linked starch with lower swelling factor was less degraded by the attack of the digestive enzymes, giving rise to RS content.

6.3.2 OSA Modification

The chemical modification of starch with octenyl succinic anhydride (OSA) is achieved by a standard esterification reaction (Bhosale & Singhal, 2006; Liu et al., 2008). Generally, OSA-starch is produced by starch and an OSA reagent in aqueous alkaline slurry system. The reaction parameters such as pH, temperature, starch concentration, weight percentage of OSA, and reaction time were proved to affect the degree of substitution (DS) and reaction efficiency (RE) of OSA-starch (Song et al., 2006; Hui et al., 2009). Modification of starch with OSA is permitted at a maximum level of 3% based on the dry starch weight basis for use in food products in many countries (CFR, 2001). Starch succinate offers a number of desirable properties such as high viscosity, better thickening power, low gelatinization, and retrogradation (Praful & Rekha, 2002). OSA starches stabilize the oil–water interface of an emulsion. The glucose part of starch binds the water and lipophilic regions; the octenyl part binds the oil, thereby preventing separation of the oil and water phases (Murphy, 2000). The optimum reaction conditions of starch esterified with alkenyl succinic anhydride are pH 8.5–9.0, reaction temperature of 23°C, and 5% anhydride concentration (Jeon et al., 1999). Esterification of starch with OSA impairs the binding of α-amylase, thus decreasing the extent of starch digestion. The research also indicated that OSA starch appeared to be a resistant starch that could be used as a functional fiber for the treatment of certain human diseases (Heacock et al., 2004). Starch succinate offers a number of desirable properties such as high viscosity, better thickening power, low gelatinization, and retrogradation (Praful & Rekha, 2002). OSA-modified starches are used in a variety of oil-in-water emulsions for food, pharmaceutical, and industrial products such as beverages and salad dressings, flavor-encapsulating agents, clouding agents, processing aids, body powders, and lotions (Jeon et al., 1999; Park et al., 2004).

Various researchers evaluated the degree of substitution (DS) of millet starches, and values of 0.20; 0.018–0.022, and 0.13, respectively, were observed (Shaikh et al., 2015; Sharma et al., 2016; Babu & Mohan, 2019). The DS of OSA-modified starch is defined as the average number of hydroxyl groups that are substituted by the OSA group per glucose unit in the starch (Tizzotti et al., 2011). Bhosale and Singhal (2006) showed that increase in DS with increase in OSA concentration could be interpreted in terms of greater availability of the OSA molecules in the proximity of the starch molecule. The DS increased as the reaction pH increased, and the optimum pH for octenyl succinylation was reported to be 8.0–8.4 (Liu et al., 2008). The DS increased when the OSA-starch was dry-heated and the pH of the starch and OSA mixture increased (Chung et al., 2010). Song et al. (2006) observed major factors that affect the esterification reaction for rice starch. It is observed that suitable parameters are selected as follows: concentration of starch slurry 35% (in proportion to water, w/w), reaction time 4h, pH of reaction system 8.5, reaction temperature 35°C, and amount of OSA 3% (in proportion to starch, w/w).

Shaikh et al. (2015) and Sharma et al. (2016) evaluated the OSA modification of pearl millet starch. It is observed that succinylated starch had higher swelling power as compared to the native counterpart. Babu and Mohan (2019) also observed similar results for OSA-modified foxtail millet starch. Punia et al. (2019) and Siroha et al. (2019a) observed similar results for mungbean and sorghum OSA-modified starches. The increase in swelling power may be due to weakening of intermolecular hydrogen bond due to introduction of bulky OSA group (Perez et al., 1993). Greater hydration and swelling of small-granule starches may be attributed to the granules having higher specific surface areas (Vasanthan & Bhatty, 1996). OSA modification shows significant effect on pasting properties and increases in peak viscosity after modification; pH of the reaction affects the peak viscosity of modified starches (Chung et al., 2010). DS directly affect the pasting properties of starches. Sharma et al. (2016) reported peak viscosity of OSA-starches of 4562–5573 mPa.s., while 2826 mPa.s. was observed for native starch. This effect is due to reaction time provided during the esterification reaction. Babu and Mohan (2019) evaluated that OSA-starch had higher peak viscosity and lesser pasting temperature as compared to native starch. Similar results are reported by Shaikh et al. (2015). This suggests that the replacement of hydroxyl groups in the starch molecules with carbonyl groups of succinic anhydride altered the granular structure of starch, thus permitting the starch granules to swell at lower temperatures and to a higher degree in comparison to native starch (Jiranuntakul et al., 2014).

Scanning electron micrograph (SEM) is used to see the effect of OSA on starch granules. Sharma et al. (2016) reported that starch granule had significant effect of OSA modification. OSA-starch granule shows slightly rough surfaces and loss of edges of granules. Some cavities and porous structures are observed, which may be due to the reaction of OSA on starch granules. Tong et al. (2019) observed similar results for OSA-modified corn starch and stated that DS affected the granules of starch. Babu and Mohan (2019) reported succinic anhydride treatment created cracks on the surface of the foxtail millet starch granule.

In the recent two decades, starch fractions have taken relevance in the food industry given the proven benefits for human health (Lehmann & Robin, 2007). Regarding this, a deep understanding of the structure of OSA-modified starches and digestibility properties is of prime importance for the design of food products with tailored digestion properties (Lopez-Silva et al., 2020). Sharma et al. (2016) evaluated higher RS content of OSA-starches as compared to native starch. Babu and Mohan (2019) reported digestibility properties of OS-modified foxtail millet starch. It is observed that after modification, RDS content decreased and the reverse is observed for RS and SDS content. Modification with OSA is known to increase levels of SDS and RS more than other modifications such as hydroxypropylation, acetylation, or cross-linking in various combinations (Han & BeMiller, 2007), resulting in high SDS functional fibers, which is considered beneficial to human nutrition (Wolf et al., 1999).

Rheological studies of OSA-starches with different OSA contents can provide valuable information on the role of OSA in native starches and also on the design of modified starches with new and improved properties (Shogren et al., 2000). Sharma et al. (2016) reported rheological properties of OSA pearl millet starch. They also observed yield stress, consistency index, and flow behavior index increased after the modification. G′, G″, and tanδ values of OSA-starch were observed to be 38.4–41.0 Pa, 6.8–11.3 Pa, and 0.18–0.29, respectively. Increased G′, G″, and tanδ values are observed for OSA-starch as compared to native starch (33.5, 4.5 Pa, and 0.14). tanδ (G″/G′) value of OSA-starch paste is observed less than 1, showing viscoelastic behavior of starch paste. A similar observation for tanδ is observed by Sandhu and Siroha (2017) and Siroha et al. (2020) for pearl millet starch.

6.3.3 Acetylation

The first research works on the possibility of starch modification *via* acetylation have appeared in 1865 (Schützenberger, 1865). In 1904, Cross and Traquair patented the method for the production of a modified preparation of acetylated starch soluble in water by heating starch with ice-cold acetic acid (Cross & Traquair, 1904). Chemical modification of starch by acetylation can be performed to significantly improve its physicochemical and functional properties. According to the Joint Food and Agriculture Organization/World Health Organization (FAO/WHO) Expert Committee on Food Additives (JECFA), starch acetate is basically defined as a starch ester prepared by addition of acetic anhydride or vinyl acetate to starch under alkaline conditions (Ali & Hasnain, 2016). The method for preparation of acetylated starch is described in Figure 6.5. Starch acetylation is a chemical modification by which part of the hydroxyl groups of glucose monomers is converted into acetyl group, altering the molecular structure of the starch. Acetylated starches are produced with acetic anhydride with an alkaline agent, such as sodium hydroxide, as catalyst (Bello-Pérez et al., 2010). During acetylation, three free hydroxyl groups on C2, C3, and C6 of the anhydroglucose unit of the starch molecule can be replaced with acetyl groups; therefore, the theoretic maximum DS is three, and the position of the OH group leads to different reactivities. The primary –OH on C6 is more reactive and is acetylated more readily than the secondary ones on C2 and C3 due to steric hindrance (Xu et al., 2004). The number of acetyl groups incorporated into the starch molecule during acetylation depend upon a number of factors such as reactant concentration, starch source (Singh et al., 2004b), reaction time, and presence of catalyst (Agboola et al., 1991). The presence of CH_3CO group yields starch with many unique properties, such as an increase in solubility, swelling power, clarity and freeze-thaw stability, reduced gelatinization temperature, and

```
┌─────────────────────────────────────┐
│              Starch                 │
└─────────────────────────────────────┘
                  ⬇
┌─────────────────────────────────────┐
│  Starch dispersed in water and pH   │
│   is adjusted 8.0 with 0.5M NaOH    │
└─────────────────────────────────────┘
                  ⬇
┌─────────────────────────────────────┐
│  Acetic anhydride is added drop-wise │
└─────────────────────────────────────┘
                  ⬇
┌─────────────────────────────────────┐
│   pH of reaction is maintained at 8.0 │
└─────────────────────────────────────┘
                  ⬇
┌─────────────────────────────────────┐
│  Reaction is continuous to 10 min with │
│               stirring              │
└─────────────────────────────────────┘
                  ⬇
┌─────────────────────────────────────┐
│   pH of reaction is adjusted at 4.5 with │
│             0.2 M HCL               │
└─────────────────────────────────────┘
                  ⬇
┌─────────────────────────────────────┐
│  Filter through 0.45μm filter membrane │
└─────────────────────────────────────┘
                  ⬇
┌─────────────────────────────────────┐
│      Oven-dried for 24 h at 40°C    │
└─────────────────────────────────────┘
                  ⬇
┌─────────────────────────────────────┐
│    Stored in air tight container    │
└─────────────────────────────────────┘
```

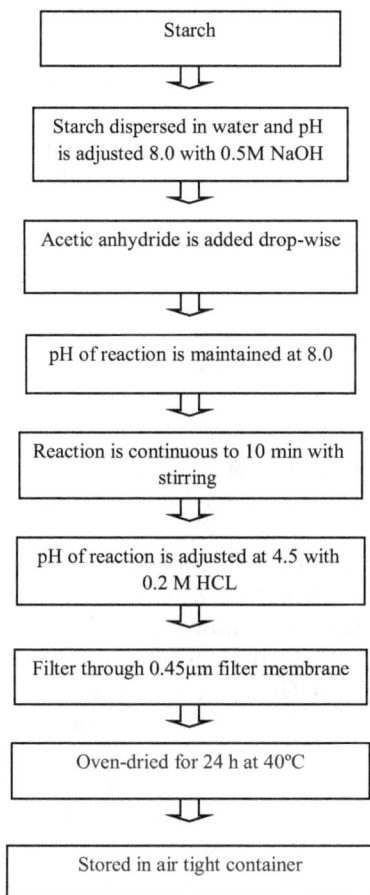

Figure 6.5   Method for preparation of acetylated starch (Wurzburg, 1964).

decreased tendency towards retrogradation (Han et al., 2012). The acetylated starches are classified on basis of DS. Acetylated starches with a low DS (0.01–0.2) may function as film-forming, binding, adhesion, thickening, stabilizing, and texturing agents. Acetylated starches with intermediate DS (0.2–1.5) and high DS (1.5–3) have high solubility in acetone and chloroform; thus, they have been reported as a thermoplastic material (Luo & Shi, 2012). Acetylated starches are extensively used in a large variety of

151

foods, including baked goods, sauces, retorted soups, frozen foods, baby foods, salad dressings, and snack foods (Wurzburg, 1995).

Siroha et al. (2019) reported acetylation modification of pearl millet starch at different levels of acetic anhydride. DS, swelling power, and solubility of acetylated starches increase as compared to native counterpart starch. Wani et al. (2012) stated that introduction of acetyl groups to starch could disorganize the intragranular structure and disrupt hydrogen bonds in the starch granules. This situation would facilitate the access of water to amorphous domains of the acetylated starches. Singh et al. (2012) and Shubeena et al. (2015) reported similar results for acetylation of sorghum and Indian horse chestnut starch. Shaikh et al. (2017) observed decrease in swelling power and solubility of acetylated pearl millet starch.

Pasting properties of modified starches altered after the inclusion of new functional group in the starch chain. Siroha et al. (2019) reported PV, BV, FV, and PT values of 1301–1772, 473–526, 1404–2120 mPa.s., and 79.4°C–84.2°C for acetylated starch, while for the native one, it was observed to be 1069, 508, 913 mPa.s., and 81.2°C, respectively. PV and FV of acetylated starch were observed to be higher than those of native starch, whereas the reverse was observed for breakdown viscosity, except starch modified at 3.75% acetic anhydride concentration. Shubeena et al. (2015) identified PV, BV, and FV increase but PT decrease is observed for acetylated Indian horse chestnut starch. Wani et al. (2012) stated an increase in viscosity when the acetyl groups are introduced into starch granules. Sun et al. (2016) observed decrease in PV, BV, FV, and PT for acetylated lotus rhizome starches. Singh et al. (2012) also observed decrease in PV, BV, and FV for acetylated sorghum starch. Thermal properties of acetylated pearl millet starch are studied by Shaikh et al. (2017). Gelatinization temperature (T_o, T_P, and T_C) of acetylated starch was observed to be 67.67°C, 70.95°C, and 75.08°C, while native starch had 69.60°C, 73.67°C, and 87.23°C, respectively. ΔH_{gel} values for native and acetylated starches were observed to be 9.86 and 2.23 (J/g). After modification, decreases in gelatinization temperature and enthalpy of gelatinization are observed. Similar results are reported by Singh et al. (2012) for sorghum starches. Lawal (2004) attributed these changes to the introduction of bulky groups in the backbone of the biopolymer, which enhances structural flexibility. Moreover, these acetyl groups may have also caused destabilization of granular structure, resulting in a decrease in gelatinization temperature (Wootton & Bamunuarachchi, 1979).

Chemical modification affects the morphology of starch granule. Siroha et al. (2019) reported that starch morphology is not affected when starch is treated with lower concentration of acetic anhydride, while higher concentration affects the morphology and cavities are observed on few starch granules. Acetylation modification shows less smooth surface with increased indentations on the surface as compared to native starch (Shaikh et al., 2019). Sodhi and Singh (2005) observed that no significant differences are seen after modification of rice starches. However, the acetylation brought about slight aggregation of granules. Singh et al. (2004a) reported that acetylation resulted in granule fusion due to its reaction with high amounts of acetic anhydride. Indeed, during the reaction, starch granules lost their integrity and were split, melted, and re-associated to form new structures (Yan & Zhengbiao, 2010).

Siroha et al. (2019) reported rheological properties of acetylated pearl millet starch. The values of G' and G" of acetylated pearl millet starches ranged between 753 and 1177 Pa and 96 and 129 Pa, respectively, during heating of starch suspension. G' value increases successfully with increase in the level of acetic anhydride. tanδ value was observed to be less than 1, resulting in viscoelastic behavior of starch suspensions. Singh et al. (2004b) observed increase in G' value after acetylation during heating for potato and corn starches. The values of G' and G" during frequency sweep test increase upon increase in angular frequency (ω); values of G' (1216–1557 Pa) were found to be much higher than G" (64.7–76.0 Pa) at all values of ω. Yield stress, consistency index, and flow behavior index were observed to be 26.6–50.7 Pa, 49.0–52.8 Pa.s^n, and 0.27–0.30, respectively, for acetylated starch. After modification, increase in yield stress and flow behavior index was observed.

Digestibility properties of acetylated pearl millet starches are reported by Siroha et al. (2019).

RDS, SDS, and RS content of acetylated starches are 44.3%–45.5%, 27.3%–32.5%, and 22.5%–27.3%, respectively. After modification, a decrease in RDS and SDS content are observed, while the reverse is observed for RS content. Ali and Hasnain (2016) stated that alpha-amylase may fail to act on this chemically altered starch structure, resulting in reduced digestibility and higher content of RS altogether.

6.3.4 Oxidation

During the oxidation process, the hydroxyl groups of the starch glucose unit were first oxidized to carbonyl groups, and then oxidized to carboxyl groups. Thereby, the degree of oxidation is indicated by the amounts of carbonyl and carboxyl groups (Huang & Lin, 2006). The common oxidants used to produce oxidized starch are permanganate (Takizawa et al., 2004), hydrogen peroxide, hypochlorite (Sangseethong et al., 2010), and oxygen (Ye et al., 2011). Hydroxyl groups, primarily at C-2, C-3, and C-6 positions, are transformed to carbonyl and/or carboxyl groups by oxidation (Kuakpetoon & Wang, 2006; Kurakake et al., 2009). Therefore, the number of these carboxyl and carbonyl groups of oxidized starch indicates the level of oxidation (Zhang et al., 2012). The food products where oxidized starch is used are neutral tasting and low viscosity such as a lemon curd, salad cream, and mayonnaise (Lawal, 2004). Oxidized starch also becomes increasingly important in the food industry for its unique functional properties such as low viscosity, high stability, clarity, film formation, and binding properties (Kuakpetoon & Wang, 2006).

Shaikh et al. (2017) observed decrease in swelling power and increase in solubility power of oxidized pearl millet starch. Hodge and Osman (1996) reported that increase in solubility after oxidation is due to disintegration and structural weakening of the starch granule. Gelatinization temperature (T_o, T_P, and T_C) and ΔH_{gel} of oxidized starch are observed to be 68.62°C, 72.92°C, and 77.50°C and 6.83 J/g, respectively. Gelatinization temperature and enthalpy of gelatinization decrease as compared to native counterpart starch. The decreases in the T_o, T_p, and T_c of starches upon oxidation might be due to the weakening of the starch granules, which led to the early rupture of the amylopectin double helices (Adebowale & Lawal, 2003). PV, hot paste viscosity, cold paste viscosity, and SV increases, while BV and PT decrease as compared to native starch (Shaikh et al., 2017).

6.3.5 Acid Treatment

In the industry, acid-modified starch is prepared with dilute hydrochloric acid (HCl) or sulfuric acid (H_2SO_4) at 25°C–55°C for various time periods (Lin et al., 2005). In acid hydrolysis, the hydroxonium ion (H_3O^+) carries out an electrophilic attack on the oxygen atom of the $\alpha(1 \rightarrow 4)$ glycosidic bond. In the next step, the electrons in one of the carbon–oxygen bonds move onto the oxygen atom to generate an unstable, high-energy carbocation intermediate. The carbocation intermediate is a Lewis acid, so it subsequently reacts with water, a Lewis base, leading to regeneration of a hydroxyl group (Hoover, 2000). Acid modification is widely used in the starch industry to produce thin boiling starches for use in food, paper, textile, and other industries (Rohwer & Klem, 1984).

Shaikh et al. (2015) reported acid thinning of pearl millet starch with 0.1 and 1.0 M HCL. Swelling power increased when starch is treated with 0.1 M HCl, while it is decreased for 1.0 M HCl-treated starch. Solubility of acid thinned starch increases as compared to native starch. Similar observations for swelling power and solubility are observed for acid modified proso millet starch (Singh &Adedeji, 2017). PV, HPV, CPV, SV, and PT of acid thinned starches decreased as compared to native counterpart starches. The highest decrease in pasting parameters is observed when starch is treated with 1.0 M HCl. Singh and Adedeji (2017) reported similar pasting properties for acid-modified starch except for PT. Gelatinization temperature (T_o, T_P, and T_C) decreases when starch is treated 0.1 M HCl, and the reverse is observed for starch treated with 1.0 M HCl and ΔH_{gel} increases after modification. Singh and Adedeji (2017) observed decrease in T_o and T_P, and increase in T_C after acid modification compared to native starch. It is suggested that the increase in gelatinization temperature could be due to the molecular structure in the starch granule (Noda et al., 1998), amylopectin chain length (Yuan et al., 1993; Jane et al., 1999), and reordering of the crystalline structure (Biliaderis et al., 1981) after hydrolysis.

6.4 ENZYMATIC MODIFICATION

Genetic and enzymatic approaches for the enhancement of food functionality have been carried out (Alexander, 1996; Davis et al., 2003). Enzyme molecules affect the granules in two ways. First, enzymes erode the outer surface of the granule and cause occurrence of characteristic fissures and pits (exocorrosion). Second, enzymes digest channels leading to the granule center weakening granule integrity and leading consequently to its breakdown (endocorrosion) (Sujka & Jamroz, 2009). Balasubramanian et al. (2014) reported decrease in swelling power and solubility of enzymatic modified pearl millet starch. PV, BV, FV, and SV decreased after modification, and the reverse was observed for PT. Enzymatic modified starches exhibited comparatively better transmittance than the native one throughout the storage period. Dey and Sit (2017) evaluated enzymatic modification of foxtail millet starch. It is observed that swelling power decreases and solubility increases after enzymatic modification. PV, HV, FV, and PT decreased as compared to native starch, while the reverse is observed for BV and SV. Texture properties of native and modified starches are also observed. Hardness and consistency decrease, while cohesiveness and index of viscosity increase as compared to native starch. Enzymatic modifications have been used to create a wide range of functional characteristics of millet starches (Table 6.3).

Table 6.3 Chemical Modification of Millet Starch

Modification	Major Results and Methodology	References
Enzymatic modification	Swelling power, solubility, PV, BD, SV, and FV decreased and PT was increased of modified starches (crude fungal amylase 0.1% was used. Starch–enzyme suspension was incubated at 37°C for 90 min in 0.04 M acetate buffer at pH 4.7). Swelling power decreases and solubility power increases after modification. PV, HV, FV, and PT decreased as compared to native starch, while the reverse is observed for BV and SV. (100 ml solution containing 10% starch w/v and 0.1% enzyme w/v was incubated with 0.04 M acetate buffer at pH 4.7 at 37°C for 90 min).	Balasubramanian et al. (2014), Dey and Sit (2017)

6.5 CONCLUSION

In this chapter, different physical, chemical, and enzymatic modifications of millet starches are discussed. Modifications alter the properties of starches due to which applications of starches increases in food and non-food applications. Food industries have shown their interest in producing more natural food components, so there is an increasing need to improve the properties of native starches without using chemical modifications, so physical modifications of starches are preferred. Chemical modification is the insertion of a new functional group on the starch backbone to give characteristic properties to the starch. The changes observed in the physicochemical, morphological, thermal, and rheological properties of the starches after modification may provide a crucial basis for understanding the efficiency of the starch modification process at industrial scale.

REFERENCES

Acierno, D., Barba, A. A. and d'Amore, M. 2004. Heat transfer phenomena during processing materials with microwave energy. *Heat and Mass Transfer* 40: 413–420.

Adebowale, K. O., Afolabi T. A. and Olu-Owolabi, B. I. 2005. Hydrothermal treatments of Finger millet (*Eleusine coracana*) starch. *Food Hydrocolloids* 19: 974–983.

Adebowale, K. O. and Lawal, O. S. 2003. Functional properties and retrogradation behaviour of native and chemically modified starch of mucuna bean (*Mucuna pruriens*). *Journal of the Science of Food and Agriculture* 83: 1541–1546.

Agboola, S. O., Akingbala, J. O. and Oguntimein, G. B. 1991. Production of low substituted cassava starch acetates and citrates. *Starch/Stärke* 43: 13–15.

Ahmed, J., Mulla, M. Z. and Arfat, Y. A. 2017. Particle size, rheological and structural properties of whole wheat flour doughs as treated by high pressure. *International Journal of Food Properties* 20: 1829–1842.

Ahmed, J., Thomas, L., Taher, A. and Joseph, A. 2016. Impact of high pressure treatment on functional, rheological, pasting, and structural properties of lentil starch dispersions. *Carbohydrate Polymers* 152: 639–647.

Alexander, R. J. 1996. New starches for food applications. *Cereal Foods World* 41: 796.

Ali, T. M. and Hasnain, A. 2016. Physicochemical, morphological, thermal, pasting, and textural properties of starch acetates. *Food Reviews International* 32: 161–180.

Amini, A. M., Razavi, S. M. A. and Mortazavi, S. A. 2015. Morphological, physicochemical, and viscoelastic properties of sonicated corn starch. *Carbohydrate Polymers* 122: 282–292.

Anderson, A. K. and Guraya, H. S. 2006. Effects of microwave heat-moisture treatment on properties of waxy and non-waxy rice starches. *Food Chemistry* 97: 318–323.

Aparicio-Saguilán, A., Gutiérrez-Meraz, F., García-Suárez, F. J., Tovar, J. and Bello-Pérez, L. A. 2008. Physicochemical and functional properties of cross-linked banana resistant starch. Effect of pressure cooking. *Starch/Stärke* 60: 286–291.

Arocas, A., Sanz, T., Hernando, M. I. and Fiszman, S. M. 2011. Comparing microwave-and water bath-thawed starch-based sauces: infrared thermography, rheology and microstructure. *Food Hydrocolloids* 25: 1554–1562.

Ashogbon, A. O. and Akintayo, E. T. 2014. Recent trend in the physical and chemical modification of starches from different botanical sources: a review. *Starch/Stärke* 66: 41–57.

Auksornsri, T., Bornhorst, E. R., Tang, J., Tang, Z. and Songsermpong, S. 2018. Developing model food systems with rice based products for microwave assisted thermal sterilization. *LWT-Food Science and Technology* 96: 551–559.

Babu, A. S. and Mohan, R. J. 2019. Influence of prior pre-treatments on molecular structure and digestibility of succinylated foxtail millet starch. *Food Chemistry* 295: 147–155.

Babu, A. S., Mohan, R. J. and Parimalavalli, R. 2019. Effect of single and dual-modifications on stability and structural characteristics of foxtail millet starch. *Food Chemistry* 271: 457–465.

Balasubramanian, S., Sharma, R., Kaur, J. and Bhardwaj, N. 2014. Characterization of modified pearl millet (*Pennisetum typhoides*) starch. *Journal of Food Science and Technology* 51: 294–300.

157

Balny, C., Masson, P. and Heremans, K. 2002. High pressure effects on biological macromolecules: from structural changes to alteration of cellular processes. *Biochimica et Biophysica Acta* 1595: 3–10.

Bao, J., Xing, J., Phillips, D. L. and Corke, H. 2003. Physical properties of octenyl succinic anhydride modified rice, wheat, and potato starches. *Journal of Agricultural and Food Chemistry* 51: 2283–2287.

Bauer, B. A. and Knorr, D. 2005. The impact of pressure, temperature and treatment time on starches: pressure-induced starch gelatinisation as pressure time temperature indicator for high hydrostatic pressure processing. *Journal of Food Engineering* 68: 329–334.

Bello-Pérez, L. A., Agama-Acevedo, E., Zamudio-Flores, P. B., Mendez-Montealvo, G. and Rodriguez-Ambriz, S. L. 2010. Effect of low and high acetylation degree in the morphological, physicochemical and structural characteristics of barley starch. *LWT-Food Science and Technology* 43: 1434–1440.

Bhosale, R. and Singhal, R. 2006. Process optimization for the synthesis of octenyl succinyl derivative of waxy corn and amaranth starches. *Carbohydrate Polymers* 66: 521–527.

Bilbao-Sáinz, C., Butler, M., Weaver, T. and Bent, J. 2007. Wheat starch gelatinization under microwave irradiation and conduction heating. *Carbohydrate Polymers* 69: 224–232.

Biliaderis, C. G., Grant, D. R. and Vose, J. R. 1981. Structural characterization of legume starches. ii: studies on acid-treated starches. *Cereal Chemistry* 58: 502–507.

Błaszczak, W., Valverde, S. and Fornal, J. 2005. Effect of high pressure on the structure of potato starch. *Carbohydrate Polymers* 59: 377–383.

Braşoveanu, M. and Nemţanu, M. R. 2014. Behaviour of starch exposed to microwave radiation treatment. *Starch/Stärke* 66: 3–14.

Bruschi, C., Komora, N., Castro, S. M., Saraiva, J., Ferreira, V. B. and Teixeira, P. 2017. High hydrostatic pressure effects on *Listeria monocytogenes* and *L. innocua*: evidence for variability in inactivation behaviour and in resistance to pediocin bacHA-6111–2. *Food Microbiology* 64: 226–231.

Carmona-Garcia, R., Sanchez-Rivera, M. M., Mendez-Montealvo, G., Garza-Montoya, B. and Bello-Perez, L. A. 2009. Effect of the cross-linked reagent type on some morphological, physicochemical and functional characteristics of banana starch (*Musa paradisiaca*). *Carbohydrate Polymers* 76: 117–122.

Castro, L. M., Alexandre, E. M., Saraiva, J. A. and Pintado, M. 2020. Impact of high pressure on starch properties: a review. *Food Hydrocolloids* 106: 105877.

CFR (Code of Federal Regulation) 2001. Food starch modified. Food Additives permitted in food for human Consumption. Govt. Printing office, Washington, USA. 21/1/172/172.892.

Chen, X., He, X., Fu, X., Zhang, B. and Huang, Q. 2017. Complexation of rice starch/flour and maize oil through heat moisture treatment: structural, *in vitro* digestion and physicochemical properties. *International Journal of Biological Macromolecules* 98: 557–564.

Chung, H. J., Lee, S. E., Han, J. A. and Lim, S. T. 2010. Physical properties of dry-heated octenyl succinylated waxy corn starches and its application in fat-reduced muffin. *Journal of Cereal Science* 52: 496–501.

Chung, H. J., Woo, K. S. and Lim, S. T. 2004. Glass transition and enthalpy relaxation of cross-linked corn starches. *Carbohydrate Polymers* 55: 9–15.

Colussi, R., Kringel, D., Kaur, L., da Rosa Zavareze, E., Dias, A. R. G. and Singh, J. 2020. Dual modification of potato starch: effects of heat-moisture and high pressure treatments on starch structure and functionalities. *Food Chemistry* 318: 126475.

Colussi, R., Singh, J., Kaur, L., da Rosa Zavareze, E., Dias, A. R. G., Stewart, R. B. and Singh, H. 2017. Microstructural characteristics and gastro-small intestinal digestion *in vitro* of potato starch: effects of refrigerated storage and reheating in microwave. *Food Chemistry* 226: 171–178.

Cross, C. F. and Traquair, J. 1904. Soluble product from starch and process of making same. U.S. Patent 778: 173.

Czechowska-Biskup, R., Rokita, B., Lotfy, S., Ulanski, P. and Rosiak, J. M. 2005. Degradation of chitosan and starch by 360-kHz ultrasound. *Carbohydrate Polymers* 60: 175–184.

Davis, J. P., Supatcharee, N., Khandelwal, R. L. and Chibbar, R. N. 2003. Synthesis of novel starches in planta: opportunities and challenges. *Starch/Stärke* 55: 107–120.

Deetae, P., Shobsngob, S., Varanyanond, W., Chinachoti, P., Naivikul, O. and Varavinit, S. 2008. Preparation, pasting properties and freeze–thaw stability of dual modified crosslink-phosphorylated rice starch. *Carbohydrate Polymers* 73: 351–358.

Deka, D. and Sit, N. 2016. Dual modification of taro starch by microwave and other heat moisture treatments. *International Journal of Biological Macromolecules* 92: 416–422.

Dey, A. and Sit, N. 2017. Modification of foxtail millet starch by combining physical, chemical and enzymatic methods. *International Journal of Biological Macromolecules* 95: 314–320.

Douzals, J. P., Perrier Cornet, J. M., Gervais, P. and Coquille, J. C. 1998. High-pressure gelatinization of wheat starch and properties of pressure-induced gels. *Journal of Agricultural and Food Chemistry* 46: 4824–4829.

Falsafi, S. R., Maghsoudlou, Y., Aalami, M., Jafari, S. M. and Raeisi, M. 2019. Physicochemical and morphological properties of resistant starch type 4 prepared under ultrasound and conventional conditions and their *in-vitro* and *in-vivo* digestibilities. *Ultrasonics Sonochemistry* 53: 110–119.

Fan, D., Gao, Y., Chen, Y., Wang, M., Gu, X., Wang, L., Shen, H., Lian, H., Zhao, J. and Zhang, H. 2017. Non-additive response of starch systems in different hydration states: a study of microwave-absorbing properties. *Innovative Food Science and Emerging Technologies* 44: 103–108.

Fan, D., Ma, W., Wang, L., Huang, J., Zhang, F., et al. 2013. Determining the effects of microwave heating on the ordered structures of rice starch by NMR. *Carbohydrate Polymers* 92: 1395–1401.

159

Fan, D., Wang, L., Shen, H., Huang, L., Zhao, J. and Zhang, H. 2017. Ultrastructure of potato starch granules as affected by microwave treatment. *International Journal of Food Properties* 20: S3189–S3194.

Flores-Silva, P. C., Roldan-Cruz, C. A., Chavez-Esquivel, G., Vernon-Carter, E. J., Bello-Perez, L. A. and Alvarez-Ramirez, J. 2017. *In vitro* digestibility of ultrasound-treated corn starch. *Starch/Stärke* 69: 1700040.

Gao, F., Li, D., Bi, C. H., Mao, Z. H. and Adhikari, B. 2014. Preparation and characterization of starch cross-linked with sodium trimetaphosphate and hydrolyzed enzymes. *Carbohydrate Polymers* 103: 310–318.

Gedanken, A. 2004. Using sonochemistry for the fabrication of nanomaterials. *Ultrasonics Sonochemistry* 11: 47–55.

Genkina, N. K., Wikman, J., Bertoft, E. and Yuryev, V. P. 2007. Effects of structural imperfection on gelatinization characteristics of amylopectin starches with A- and B-type crystallinity. *Biomacromolecules* 8: 2329–2335.

Gomes, M. R., Clark, A. and Ledward, D. A. 1998. Effects of high pressure on amylases and starch in wheat and barley flours. *Food Chemistry* 63: 363–372.

Gunaratne, A. and Hoover, R. 2002. Effect of heat–moisture treatment on the structure and physicochemical properties of tuber and root starches. *Carbohydrate Polymers* 49: 425–437.

Han, J. A. and BeMiller, J. N. 2007. Preparation and physical characteristics of slowly digesting modified food starches. *Carbohydrate Polymers* 67: 366–374.

Han, F., Liu, M., Gong, H., Lü, S., Ni, B. and Zhang, B. 2012. Synthesis, characterization and functional properties of low substituted acetylated corn starch. *International Journal of Biological Macromolecules* 50: 1026–1034.

Heacock, P. M., Hertzler, S. R. and Wolf, B. 2004. The glycemic, insulinemic, and breath hydrogen responses in humans to a food starch esterified by 1-octenyl succinic anhydride. *Nutrition Research* 24: 581–592.

Hirsch, J. B. and Kokini, J. L. 2002. Understanding the mechanism of cross-linking agents (POCl$_3$, STMP, and EPI) through swelling behavior and pasting properties of cross-linked waxy maize starches. *Cereal Chemistry* 79: 102–107.

Hodge, J. E. and Osman, E. M. 1996. Carbohydrates. In Fennema, O. R., editor. *Food chemistry.* New York: Marcel Dekker.

Hong, J. S., Gomand, S. V. and Delcour, J. A. 2015. Preparation of cross-linked maize (*Zea mays* L.) starch in different reaction media. *Carbohydrate Polymers* 124, 302–310.

Hoover, R. 2000. Acid-treated starches. *Food Reviews International* 16: 369–392.

Hoover, R. 2010. The impact of heat-moisture treatment on molecular structures and properties of starches isolated from different botanical sources. *Critical Reviews in Food Science and Nutrition* 50: 835–847.

Hoover, R. and Vasanthan, T. 1994. The flow properties of native, heat-moisture treated, and annealed starches from wheat, oat, potato and lentil. *Journal of Food Biochemistry* 18: 67–82.

Huang, L. F. X. F. X. and Lin, Q. L. 2006. Properties and cross linking mechanism of starches oxidized by sodium hypochlorite at low level. *Journal of South China University of Technology (Natural Science Edition)* 34: 79–83.

Huang, H. W., Wu, S. J., Lu, J. K., Shyu, Y. T. and Wang, C. Y. 2017. Current status and future trends of high-pressure processing in food industry. *Food Control* 72: 1–8.

Huber, K. C. and BeMiller, J. N. 2001. Location of sites of reaction within starch granules. *Cereal Chemistry* 78(2): 173–180.

Hui, R., Chen, Q.-H., Fu, M.-L., Xu, Q. and He, G. Q. 2009. Preparation and properties of octenyl succinc anhydride modified potato starch. *Food Chemistry* 114: 81–86.

Jambrak, A. R., Herceg, Z., Šubarić, D., Babić, J., Brnčić, M., et al. 2010. Ultrasound effect on physical properties of corn starch. *Carbohydrate Polymers* 79: 91–100.

Jane, J. L., Chen, Y. Y., Lee, L. F., McPherson, A. E., Wong, K. S., Radosavljevic, M. and Kasemsuwan, T. 1999. Effects of amylopectin branch chain length and amylose content on the gelatinization and pasting properties of starch. *Cereal Chemistry* 76: 629–637.

Jayakody, L. and Hoover, R. 2002. The effect of linterization on cereal starch granules. *Food Research International* 35: 665–680.

Jayakody, L. and Hoover, R. 2008. Effect of annealing on the molecular structure and physicochemical properties of starches from different botanical origins–a review. *Carbohydrate Polymers* 74: 691–703.

Jeon, Y., Lowell, A. V. and Gross, R. A. 1999. Studies of starch esterification: reactions with alkenyl succinates in aqueous slurry systems. *Starch/Stärke* 51: 90–93.

Jiranuntakul, W., Puncha-arnon, S. and Uttapap, D. 2014. Enhancement of octenyl succinylation of cassava starch by prior modification with heat-moisture treatment. *Starch/Stärke* 66: 1071–1078.

Jyothi, A. N., Moorthy, S. N. and Rajasekharan, V. 2006. Effect of cross linking with epichlorohydrin on the properties of cassava (*Manihot esculenta Crantz*) starch. *Starch/Stärke* 58: 292–299.

Kaur, L., Singh, J. and Singh, N. 2006. Effect of cross-linking on some properties of potato (*Solanum tuberosum* L.) starches. *Journal of the Science of Food and Agriculture* 86: 1945–1954.

Kavitha, R. and BeMiller, J. N. 1998. Characterization of hydroxypropylated potato starch. *Carbohydrate Polymers* 37: 115–121.

Kawai, K., Fukami, K. and Yamamoto, K. 2012. Effect of temperature on gelatinization and retrogradation in high hydrostatic pressure treatment of potato starch–water mixtures. *Carbohydrate Polymers* 87: 314–321.

Kim, H., Kim, B. Y. and Baik, M. Y. 2012. Application of ultra high pressure (UHP) in starch chemistry. *Critical Reviews in Food Science and Nutrition* 52: 123–141.

Kim, B. Y. and Yoo, B. 2010. Effects of cross-linking on the rheological and thermal properties of sweet potato starch. *Starch/Stärke* 62: 577–583.

161

Kittipongpatana, O. S. and Kittipongpatana, N. 2013. Physicochemical, in vitro digestibility and functional properties of carboxymethyl rice starch cross-linked with epichlorohydrin. *Food Chemistry* 141: 1438–1444.

Koo, S. H., Lee, K. Y. and Lee, H. G. 2010. Effect of cross-linking on the physico-chemical and physiological properties of corn starch. *Food Hydrocolloids* 24: 619–625.

Krueger, B. R., Walker, C. E., Knutson, C. A. and Inglett, G. E. 1987. Differential scanning calorimetry of raw and annealed starch isolated from normal and mutant maize genotypes. *Cereal Chemistry* 64: 187–190.

Kuakpetoon, D. and Wang, Y. 2006. Structural characteristics and physico-chemical properties of oxidized corn starches varying in amylose content. *Carbohydrate Research* 341: 1896–1915.

Kumar, Y., Singh, L., Sharanagat, V. S., Patel, A. and Kumar, K. 2020. Effect of microwave treatment (low power and varying time) on potato starch: microstructure, thermo-functional, pasting and rheological properties. *International Journal of Biological Macromolecules* 155: 27–35.

Kurakake, M., Akiyama, Y., Hagiwara, H. and Komaki, T. 2009. Effects of cross-linking and low molecular amylose on pasting characteristics of waxy corn starch. *Food Chemistry* 116: 66–70.

Lawal, O. S. 2004. Composition, physicochemical properties and retrograda-tion characteristics of native, oxidized, acetylated and acid-thinned new cocoyam (*Xanthosoma sagittifolium*) starch. *Food Chemistry* 87: 205–218.

Lehmann, U. and Robin, F. 2007. Slowly digestible starch–its structure and health implications: a review. *Trends in Food Science and Technology* 18: 346–355.

Li, W., Bai, Y., Mousaa, S. A. S., Zhang, Q. and Shen, Q. 2012. Effect of high hydro-static pressure on physicochemical and structural properties of rice starch. *Food and Bioprocess Technology* 5: 2233–2241.

Li, W., Gao, J., Saleh, A. S., Tian, X., Wang, P., Jiang, H. and Zhang, G. 2018. The modifications in physicochemical and functional properties of proso millet starch after ultra-high pressure (UHP) process. *Starch/Stärke* 70: 1700235.

Li, Y., Hu, A., Wang, X. and Zheng, J. 2019a. Physicochemical and in vitro diges-tion of millet starch: effect of moisture content in microwave. *International Journal of Biological Macromolecules* 134: 308–315.

Li, Y., Hu, A., Zheng, J. and Wang, X. 2019b. Comparative studies on structure and physiochemical changes of millet starch under microwave and ultra-sound at the same power. *International Journal of Biological Macromolecules* 141: 76–84.

Li, B. Z., Wang, L. J., Chiu, Y. L., Zhang, Z. J., Shi, J., et al. 2009. Physical proper-ties and loading capacity of starch-based microparticles crosslinked with trisodium trimetaphosphate. *Journal of Food Engineering* 92: 255–260.

Li, S., Ward, R. and Gao, Q. 2011. Effect of heat-moisture treatment on the formation and physicochemical properties of resistant starch from mung bean (*Phaseolus radiatus*) starch. *Food Hydrocolloids* 25: 1702–1709.

Lim, S. T. and Seib, P. A. 1993. Location of phosphate esters in a wheat starch phosphate by[31]P-nuclear magnetic resonance spectroscopy. *Cereal Chemistry* 70: 145–152.

Lin, J. H., Lii, C. Y. and Chang, Y. H. 2005. Change of granular and molecular structures of waxy maize and potato starches after treated in alcohols with or without hydrochloric acid. *Carbohydrate Polymers* 59: 507–515.

Lin, C. L., Lin, J. H., Lin, J. J. and Chang, Y. H. 2019. Progressive alterations in crystalline structure of starches during heat-moisture treatment with varying iterations and holding times. *International Journal of Biological Macromolecules* 135: 472–480.

Liu, Z., Li, Y., Cui, F., Ping, L., Song, J., et al. 2008. Production of octenyl succinic anhydride-modified waxy corn starch and its characterization. *Journal of Agricultural and Food Chemistry* 56: 11499–11506.

Liu, K., Zhang, B., Chen, L., Li, X. and Zheng, B. 2019. Hierarchical structure and physicochemical properties of highland barley starch following heat moisture treatment. *Food Chemistry* 271: 102–108.

Lopez-Silva, M., Bello-Perez, L. A., Castillo-Rodriguez, V. M., Agama-Acevedo, E. and Alvarez-Ramirez, J. 2020. In vitro digestibility characteristics of octenyl succinic acid (OSA) modified starch with different amylose content. *Food Chemistry* 304: 125434.

Lorimer, J. P. and Mason, T. J. 1987. Sonochemistry. Part 1-the physical aspects. *Chemical Society Reviews* 16: 239–274.

Lu, Z. H., Belanger, N., Donner, E. and Liu, Q. 2018. Debranching of pea starch using pullulanase and ultrasonication synergistically to enhance slowly digestible and resistant starch. *Food Chemistry* 268: 533–541.

Luo, Z., Fu, X., He, X., Luo, F., Gao, Q. and Yu, S. 2008. Effect of ultrasonic treatment on the physicochemical properties of maize starches differing in amylose content. *Starch/Stärke* 60: 646–653.

Luo, Z. G., He, X., Fu, X., Luo, F. X. and Gao, Q. 2006. Effect of microwave radiation on the physicochemical properties of normal maize, waxy maize and amylomaize starches. *Starch/Stärke* 58: 468–474.

Luo, Z. G. and Shi, Y. C. 2012. Preparation of acetylated waxy, normal, and high-amylose maize starches with intermediate degrees of substitution in aqueous solution and their properties. *Journal of Agricultural and Food Chemistry* 60: 9468–9475.

Ma, S., Fan, D., Wang, L., Lian, H., Zhao, J., Zhang, H. and Chen, W. 2015. The impact of microwave heating on the granule state and thermal properties of potato starch. *Starch/Stärke* 67: 391–398.

Maache-Rezzoug, Z., Zarguili, I., Loisel, C., Queveau, D. and Buleon, A. 2008. Structural modifications and thermal transitions of standard maize starch after DIC hydrothermal treatment. *Carbohydrate Polymers* 74: 802–812.

Majzoobi, M., Seifzadeh, N., Farahnaky, A. and Mesbahi, G. 2015. Effects of sonication on physical properties of native and cross-linked wheat starches. *Journal of Texture Studies* 46: 105–112.

163

Mehboob, S., Ali, T. M., Sheikh, M. and Hasnain, A. 2020. Effects of cross linking and/or acetylation on sorghum starch and film characteristics. *International Journal of Biological Macromolecules* 155: 786–794.

Mollekopf, N., Treppe, K., Fiala, P. and Dixit, O. 2011. Vacuum microwave treatment of potato starch and the resultant modification of properties. *Chemie Ingenieur Technik* 83: 262–272.

Monroy, Y., Rivero, S. and García, M. A. 2018. Microstructural and techno-functional properties of cassava starch modified by ultrasound. *Ultrasonics Sonochemistry* 42: 795–804.

Murphy, P. 2000. Starch. In: Phillips, G. O. and Williams, editors. *Handbook of hydrocolloids*. Boca Raton, FL: CRC Press. pp. 41–65.

Nadir, A. S., Helmy, I. M. F., Nahed, M. A., Wafaa, M. M. A. and Ramadan, M. T. 2015. Modification of potato starch by some different physical methods and utilization in cookies production. *International Journal of Current Microbiology and Applied Sciences* 4: 556–569.

Nakazawa, Y. and Wang, Y. J. 2003. Acid hydrolysis of native and annealed starches and branch-structure of their Naegeli dextrins. *Carbohydrate Research* 338: 2871–2882.

Nakazawa, Y. and Wang, Y. J. 2004. Effect of annealing on starch–palmitic acid interaction. *Carbohydrate Polymers* 57: 327–335.

Noda, T., Takahata, Y., Sato, T., Suda, I., Morishita, T., Ishiguro, K. and Yamakawa, O. 1998. Relationships between chain length distribution of amylopectin and gelatinization properties within the same botanical origin for sweet potato and buckwheat. *Carbohydrate Polymers* 37: 153–158.

Olayinka, O. O., Adebowale, K. O. and Olu-Owolabi, B. I. 2008. Effect of heat-moisture treatment on physicochemical properties of white sorghum starch. *Food Hydrocolloids* 22: 225–230.

Ortega-Ojeda, F. E. and Eliasson, A. C. 2001. Gelatinisation and retrogradation behaviour of some starch mixtures. *Starch/Stärke* 53: 520–529.

Oyeyinka, S. A., Umaru, E., Olatunde, S. J. and Joseph, J. K. 2019. Effect of short microwave heating time on physicochemical and functional properties of Bambara groundnut starch. *Food Bioscience* 28: 36–41.

Park, S., Chung, M. and Yoo, B. 2004. Effect of octenyl succinylation on rheological properties of corn starch pastes. *Starch/Stärke* 56: 399–406.

Patist, A. and Bates, D. 2008. Ultrasonic innovations in the food industry: from the laboratory to commercial production. *Innovative Food Science and Emerging Technologies* 9: 147–154.

Perez, E., Bahanassey, Y. A. and Breene, W. M. 1993. Some chemical, physical and functional properties of native and modified starches of *Amaranthus hypochondiracus* and *Amaranthus cruentus*. *Starch/Stärke* 45: 215–220.

Praful, N. B. and Rekha, S. S. 2002. Effect of succinylation on the corn and amaranth starch pastes. *Carbohydrate Polymers* 48: 233–240.

Punia, S. 2020. Barley starch modifications: physical, chemical and enzymatic–a review. *International Journal of Biological Macromolecules* 144: 578–585.

164

Punia, S., Siroha, A. K., Sandhu, K. S. and Kaur, M. 2019. Rheological and pasting behavior of OSA modified mungbean starches and its utilization in cake formulation as fat replacer. *International Journal of Biological Macromolecule* 128: 230–236.

Ratnayake, W. S. and Jackson, D. S. 2008. Phase transition of cross-linked and hydroxypropylated corn (*Zea mays* L.) starches. *LWT-Food Science and Technology* 41: 346–358.

Rohwer, R. G. and Klem, R. E. 1984. Acid-modified starch: production and use. In. Bemiller, J. N. and Paschall, E. F., editors. *Starch: chemistry and technology.* Orlando, FL: Academic Press. pp. 529–541.

Salazar-González, C., San Martín-González, M. F., López-Malo, A. and Sosa-Morales, M. E. 2012. Recent studies related to microwave processing of fluid foods. *Food and Bioprocess Technology* 5: 31–46.

Sandhu, K. S. and Siroha, A. K. 2017. Relationships between physicochemical, thermal, rheological and in vitro digestibility properties of starches from pearl millet cultivars. *LWT-Food Science and Technology* 83: 213–224.

Sandhu, K. S., Siroha, A. K., Punia, S. and Nehra, M. 2020. Effect of heat moisture treatment on rheological and in vitro digestibility properties of pearl millet starches. *Carbohydrate Polymer Technologies and Applications* 1: 100002.

Sangseethong, K., Termvejsayanon, N. and Sriroth, K. 2010. Characterization of physicochemical properties of hypochlorite- and peroxide-oxidized cassava starches. *Carbohydrate Polymers* 82: 446–453.

San Martin, M. F., Barbosa-Cánovas, G. V. and Swanson, B. G. 2002. Food processing by high hydrostatic pressure. *Critical Reviews in Food Science and Nutrition* 42: 627–645.

Schützenberger, P. 1865. Action de l'acide acétique anhydre sur la cellulose, l'amidon, les sucres, la mannite et ses congénères, les glucosides et certaines matières colorants végétales. *Comptes rendus de l'Académie des Sciences* 61: 484–487.

Seguchi, M. and Hignsa, T. 1994. Study of wheat starch structures by sonication treatment. *Cereal Chemistry* 71: 639–641.

Shah, U., Gani, A., Ashwar, B. A., Shah, A., Wani, I. A. and Masoodi, F. A. 2016. Effect of infrared and microwave radiations on properties of Indian Horse Chestnut starch. *International Journal of Biological Macromolecules* 84: 166–173.

Shaikh, M., Ali, T. M. and Hasnain, A. 2015. Post succinylation effects on morphological, functional and textural characteristics of acid-thinned pearl millet starches. *Journal of Cereal Science* 63: 57–63.

Shaikh, M., Ali, T. M. and Hasnain, A. 2017. Utilization of chemically modified pearl millet starches in preparation of custards with improved cold storage stability. *International Journal of Biological Macromolecules* 104: 360–366.

Shaikh, M., Haider, S., Ali, T. M. and Hasnain, A. 2019. Physical, thermal, mechanical and barrier properties of pearl millet starch films as affected by levels of acetylation and hydroxypropylation. *International Journal of Biological Macromolecules* 124: 209–219.

165

Sharma, M., Singh, A. K., Yadav, D. N., Arora, S. and Vishwakarma, R. K. 2016. Impact of octenyl succinylation on rheological, pasting, thermal and physicochemical properties of pearl millet (*Pennisetum typhoides*) starch. *LWT-Food Science and Technology* 73: 52–59.

Sharma, M., Yadav, D. N., Singh, A. K. and Tomar, S. K. 2015. Rheological and functional properties of heat moisture treated pearl millet starch. *Journal of Food Science and Technology* 52: 6502–6510.

Shen, H., Fan, D., Huang, L., Gao, Y., Lian, H., Zhao, J. and Zhang, H. 2017. Effects of microwaves on molecular arrangements in potato starch. *RSC Advances* 7: 14348–14353.

Shi, Y. C. 2008. Two-and multi-step annealing of cereal starches in relation to gelatinization. *Journal of Agricultural and Food Chemistry* 56(3): 1097–1104.

Shogren, R. L., Viswanathan, A., Felker, F. and Gross, R. A. 2000. Distribution of octenyl succinate groups in octenyl succinic anhydride modified waxy maize starch. *Starch/Stärke* 52: 196–204.

Shubeena, Wani, I. A., Gani, A., Sharma, P., Wani, T. A., Masoodi, F. A., Hamdani, A. and Muzafar, S. 2015. Effect of acetylation on the physico-chemical properties of Indian Horse Chestnut (*Aesculus indica* L.) starch. *Starch/Stärke* 67: 311–318.

Singh, M. and Adedeji, A. A. 2017. Characterization of hydrothermal and acid modified proso millet starch. *LWT-Food Science and Technology* 79: 21–26.

Singh, N., Chawla, D. and Singh, J. 2004a. Influence of acetic anhydride on physicochemical, morphological and thermal properties of corn and potato starch. *Food Chemistry* 86: 601–608.

Singh, J., Kaur, L. and McCarthy, O. J. 2007. Factors influencing the physicochemical morphological thermal and rheological properties of some chemically modified starches for food application–a review. *Food Hydrocolloids* 21: 1–22.

Singh, J., Kaur, L. and Singh, N. 2004b. Effect of acetylation on some properties of corn and potato starches. *Starch/Stärke* 56: 586–601.

Singh, H., Sodhi, N. S. and Singh, N. 2012. Structure and functional properties of acetylated sorghum starch. *International Journal of Food Properties* 15: 312–325.

Siroha, A. K., Punia, S., Sandhu, K. S. and Karwasra, B. L. 2020. Physicochemical, pasting, and rheological properties of pearl millet starches from different cultivars and their relations. *Acta Alimentaria* 49: 49–59.

Siroha, A. K. and Sandhu, K. S. 2018. Physicochemical, rheological, morphological, and in vitro digestibility properties of cross-linked starch from pearl millet cultivars. *International Journal of Food Properties* 21: 1371–1385.

Siroha, A. K., Sandhu, K. S., Kaur, M. and Kaur, V. 2019. Physicochemical, rheological, morphological and in vitro digestibility properties of pearl millet starch modified at varying levels of acetylation. *International Journal of Biological Macromolecules* 131: 1077–1083.

Siroha, A. K., Sandhu, K. S. and Punia, S. 2019a. Impact of octenyl succinic anhydride on rheological properties of sorghum starch. *Quality Assurance and Safety of Crops & Foods* 11: 221–229.

Sodhi, N. S. and Singh, N. 2005. Characteristics of acetylated starches prepared using starches separated from different rice cultivars. *Journal of Food Engineering* 70: 117–127.

Solarek, D. B. 1986. *Modified starches: properties and uses*. Boca Raton, FL: CRC Press. pp. 97–112.

Song, X. Y., He, G. Q., Ruan, H. and Chen, Q. H. 2006. Preparation and properties of octenyl succinic anhydride modified early indica rice starch. *Starch/Stärke* 58: 109–117.

Stolt, M., Oinonen, S. and Autio, K. 2000. Effect of high pressure on the physical properties of barley starch. *Innovative Food Science and Emerging Technologies* 1: 167–175.

Sujka, M. and Jamroz, J. 2009. α-Amylolysis of native potato and corn starches– SEM, AFM, nitrogen and iodine sorption investigations. *LWT-Food Science and Technology* 42: 1219–1224.

Sujka, M. and Jamroz, J. 2013. Ultrasound-treated starch: SEM and TEM imaging, and functional behaviour. *Food Hydrocolloids* 31: 413–419.

Sun, S., Zhang, G. and Ma, C. 2016. Preparation, physicochemical characterization and application of acetylated lotus rhizome starches. *Carbohydrate Polymers* 135: 10–17.

Takizawa, F. F., Silva, G. O., Konkel, F. E. and Demiate, I. M. 2004. Characterization of tropical starches modified with potassium permanganate and lactic acid. *Brazilian Archives of Biology and Technology* 47: 921–931.

Tan, X., Li, X., Chen, L., Xie, F., Li, L. and Huang, J. 2017. Effect of heat-moisture treatment on multi-scale structures and physicochemical properties of breadfruit starch. *Carbohydrate Polymers* 161: 286–294.

Tester, R. F. and Debon, S. J. 2000. Annealing of starch–a review. *International Journal of Biological Macromolecules* 27: 1–12.

Tizzotti, M. J., Sweedman, M. C., Tang, D., Schaefer, C. and Gilbert, R. G. 2011. New 1H NMR procedure for the characterization of native and modified food-grade starches. *Journal of Agricultural and Food Chemistry* 59: 6913–6919.

Tokuşoğlu, Ö. and Swanson, B. G. 2014. Introduction to improving food quality by novel food processing. In: Tokuşoğlu, O. and Swanson, B. G., editor. *Improving food quality with novel food processing technologies*. London: CRC Press. pp. 16–21.

Tomasik, P. and Zaranyika, M. F. 1995. Nonconventional methods of modification of starch. In: Horton, D., editor. *Advances in carbohydrate chemistry and biochemistry*. Vol. 51. US: Academic Press. pp. 243–318.

Tong, F., Deng, L., Sun, R. and Zhong, G. 2019. Effect of octenyl succinic anhydride starch ester by semi-dry method with vacuum-microwave assistant. *International Journal of Biological Macromolecules* 141: 1128–1136.

Vasanthan, T. and Bhatty, R. S. 1996. Physicochemical properties of small and large granule starches of waxy, regular, and high amylose barleys. *Cereal Chemistry* 73: 199–207.

Vermeylen, R., Goderis, B. and Delcour, J. A. 2006. An X-ray study of hydrothermally treated potato starch. *Carbohydrate Polymers* 64: 364–375.

Wang, S., Jin, F. and Yu, J. 2013. Pea starch annealing: new insights. *Food and Bioprocess Technology* 6: 3564–3575.

Wang, H., Wang, Z., Li, X., Chen, L. and Zhang, B. 2017. Multi-scale structure, pasting and digestibility of heat moisture treated red adzuki bean starch. *International Journal of Biological Macromolecules* 102: 162–169.

Wani, I. A., Sogi, D. S. and Gill, B. S. 2012. Physicochemical properties of acetylated starches from some Indian kidney bean (*Phaseolus vulgaris* L.) cultivars. *International Journal of Food Science and Technology* 47: 1993–1999.

Watcharatewinkul, Y., Puttanlek, C., Rungsardthong, V. and Uttapap, D. 2009. Pasting properties of a heat-moisture treated canna starch in relation to its structural characteristics. *Carbohydrate Polymers* 75: 505–511.

Wolf, B. W., Bauer, L. L. and Fahey, G. C. 1999. Effects of chemical modification on in vitro rate and extent of food starch digestion: an attempt to discover a slowly digested starch. *Journal of Agricultural and Food Chemistry* 47: 4178–4183.

Woo, K. S. 1999. Cross-linked, RS4 type resistant starch: preparation and properties. PhD thesis. Kansas State: University Manhattan, KS.

Woo, K. S. and Seib, P. A. 1997. Cross-linking of wheat starch and hydroxypropylated wheat starch in alkaline slurry with sodium trimetaphosphate. *Carbohydrate Polymers* 33: 263–271.

Wootton, M. and Bamunuarachchi, A. 1979. Application of differential scanning calorimetry to starch gelatinization. I. Commercial native and modified starches. *Starch/Stärke* 31: 201–204.

Wurtzburg, O. B. 1960. Preparation of starch derivatives. US Patent No. 2: 935,510.

Wurzburg, O. B. 1964. Starch derivatives and modification. In: Whistler, R. L., editor. *Methods in carbohydrate chemistry*. 4th ed. New York: Academic Press. pp. 286–288.

Wurzburg, O. B. 1986. Cross-linked starches. In Wurzburg, O. B., editor. *Modified starches: properties and uses*. Florida: CRC Press. p. 41e53.

Wurzburg, O. B. 1995. Modified starches. In: Stephen, A. M., editor. *Food polysaccharides and their applications*. New York: Marcel Dekker. pp. 83–85.

Xia, T., Gou, M., Zhang, G., Li, W. and Jiang, H. 2018. Physical and structural properties of potato starch modified by dielectric treatment with different moisture content. *International Journal of Biological Macromolecules* 118: 1455–1462.

Xie, X. and Liu, Q. 2004. Development and physicochemical characterization of new resistant citrate starch from different corn starches. *Starch/Stärke* 56: 364–370.

Xie, Y., Yan, M., Yuan, S., Sun, S. and Huo, Q. 2013. Effect of microwave treatment on the physicochemical properties of potato starch granules. *Chemistry Central Journal* 7: 1–7.

Xu, Y., Miladinov, V. and Hanna, M. A. 2004. Synthesis and characterization of starch acetates with high substitution. *Cereal Chemistry* 81: 735–740.

Yan, H. and Zhengbiao, G. U. 2010. Morphology of modified starches prepared by different methods. *Food Research International* 43: 767–772.

Yang, W., Kong, X., Zheng, Y., Sun, W., Chen, S., et al. 2019a. Controlled ultrasound treatments modify the morphology and physical properties of rice starch rather than the fine structure. *Ultrasonics Sonochemistry* 59: 104709.

Yang, Q. Y., Lu, X. X., Chen, Y. Z., Luo, Z. G. and Xiao, Z. G. 2019b. Fine structure, crystalline and physicochemical properties of waxy corn starch treated by ultrasound irradiation. *Ultrasonics Sonochemistry* 51: 350–358.

Ye, S., Wang, Q. H., Xu, X. C., Jiang, W. Y., Gan, S. C. and Zou, H. F. 2011. Oxidation of cornstarch using oxygen as oxidant without catalyst. *LWT-Food Science and Technology* 44: 139–144.

Yuan, R. C., Thompson, D. B. and Boyer, C. D. 1993. Fine structure of amylopectin in relation to gelatinization and retrogradation behavior of maize starches from three wx-containing genotypes in two inbred lines. *Cereal Chemistry* 70: 81–89.

Zeng, F., Ma, F., Kong, F., Gao, Q. and Yu, S. 2015. Physicochemical properties and digestibility of hydrothermally treated waxy rice starch. *Food Chemistry* 172: 92–98.

Zhang, Y. R., Wang, X. L., Zhao, G. M. and Wang, Y. Z. 2012. Preparation and properties of oxidized starch with high degree of oxidation. *Carbohydrate Polymers* 87: 2554–2562.

Zheng, J., Li, Q., Hu, A., Yang, L., Lu, J., Zhang, X. and Lin, Q. 2013. Dual-frequency ultrasound effect on structure and properties of sweet potato starch. *Starch/Stärke* 65: 621–627.

Zheng, M. Z., Xiao, Y., Yang, S., Liu, H. M., Liu, M. H., et al. 2020. Effects of heat–moisture, autoclaving, and microwave treatments on physicochemical properties of proso millet starch. *Food Science and Nutrition* 8: 735–743.

Zhu, F. 2015. Impact of ultrasound on structure, physicochemical properties, modifications, and applications of starch. *Trends in Food Science and Technology* 43: 1–17.

Zhu, J., Li, L., Chen, L. and Li, X. 2012. Study on supramolecular structural changes of ultrasonic treated potato starch granules. *Food Hydrocolloids* 29: 116–122.

Zia-ud-Din, Xiong, H. and Fei, P. 2017. Physical and chemical modification of starches: a review. *Critical Reviews in Food Science and Nutrition* 57: 2691–2705.

Zuo, J. Y., Knoerzer, K., Mawson, R., Kentish, S. and Ashokkumar, M. 2009. The pasting properties of sonicated waxy rice starch suspensions. *Ultrasonics Sonochemistry* 16: 462–468.

7

Bioactive Compounds of Millets

7.1 INTRODUCTION

Since human civilization began, cereal grains have been preferred as nutrient-rich edible sources that are further divided into family, genus, order, etc. on the basis of their botanical characteristics. The grains are agro-industrially important substrates as they could be utilized as food/feed and for the preparation of health benefiting products. Worldwide, many cereal crops are being grown by agricultural sector; however, among them millets have their own importance due to the presence of minerals, starch, fibers, protein, and health-benefiting antioxidant compounds (Kaur et al., 2019; Shahidi & Chandrasekara, 2013; Chandrasekara and Shahidi, 2011a). Being a grass family member, millets are also agriculturally important as they have the capability to tolerate harsher environmental conditions and can grow under minimal soil moisture content (Sandhu et al., 2020; Siroha & Sandhu, 2017). These characteristic features of millets help the farmers to achieve sustainability as they have an optional crop that can grow within a limited range of moisture and soil nutrients. Further, the gluten-free nature of millets makes them an industrially important crop for the preparation of various bakery products from them (Niro et al., 2019; Brasil et al., 2015; Taylor & Emmambux, 2008).

Bioactive compounds are considered as important metabolites that have health-benefiting features in them ranging from antimicrobial potential, anti-cancerous properties, and anti-aging to DNA damage protection potential (Purewal et al., 2019; Sandhu & Punia, 2017;

Siroha et al., 2016). Documented reports on antioxidants demonstrated that the health benefiting potential of these secondary metabolites are due to the presence of a unique blend of specific compounds (Fardet, 2010). Millets' potential against diabetes, cholesterol, DNA damage, and cancer has been explored by many scholars/scientists all over the World (Sharma & Niranjan, 2017; Bangoura et al., 2013; Ju-Sung et al., 2011; Sireesha et al., 2011; Lee et al., 2010; Anderson et al., 2009; Choi et al., 2005). Like other major cereal grains, millets also have phenolic compounds in them; however, a significant proportion of them are present in bound form, which under certain sets of conditions may be released into free form. These parameters include selection of appropriate extraction medium (solvent concentration, type, temperature range, and time duration) (Dzah et al., 2020; Canadas et al., 2020; Skoronski et al., 2020; Irakli et al., 2012; Dai & Mumper, 2010). In addition to these factors, one more important thing to achieve maximum experimental output is the use of statistical software. Response surface methodology (RSM) is widely being used to enhance the overall output from the experimental setup (Singh et al., 2019; Sharma et al., 2018; Bhatti et al., 2016; Dureja et al., 2014; Kaith et al., 2012). This chapter helps the readers to understand the various types of bioactive compounds present in millets. Further, the chapter also discusses the factors that affects the recovery of phenolics from millets.

7.2 BIOACTIVE COMPOUNDS AND OXIDATIVE STRESS

Metabolites that either possesses any biological activity or directly affect the health of living organisms could be referred to as bioactive compounds/constituent. The number of synthetic antioxidant-rich products available in the market for consumers is large; however, they are not considered as safe as natural products. This situation may raise the necessity for products of natural origin. Grains are being studied to explore their health-benefiting components. Nutritional as well as bioactive profile of millets indicates that they could be used as natural medicine. Natural resources have antioxidant-rich compounds as secondary metabolites, which may form/be synthesized in response to physiological/ecological pressure as well as due to their exposure to UV radiations, attack by pathogens, wounding, and limited availability of nutrients (Dhull et al., 2020; Kaur et al., 2018; Salar et al., 2015; Zulak et al., 2006; Zieliński & Kozłowska, 2000). Further, the concentration and type of specific phenolics in natural resources may vary with the quantification methods adopted, cultivar type, soil conditions (temperature, nutritional profile, pH, and rainfall), and storage conditions (Salar et al., 2017a; Adom & Liu, 2002; Adom et al., 2005).

OS is the process during which an imbalance occurs between bioactive compounds and reactive oxygen species. Oxidants may be produced in cellular systems by many enzymes that utilize molecular oxygen as a substrate (Finkel, 2003). OS involves conditions during which free radicals (FRs) deploy their harmful effects due to their increased production and changed cellular mechanisms. Due to high concentration of unsaturated fatty acids (UFAs), cellular membranes become vulnerable to oxidation. FRs are characterized by the presence of unpaired electrons and have potential to exist independently. Further, FRs are capable enough to start their damaging effects due to potential binding with cellular components (Salar et al., 2017a). Willcox et al. (2004) demonstrated that the presence of unpaired electrons is the sole reason that makes FRs more reactive and less stable. FRs react with cellular components and make damaging complexes with proteins, lipids, and DNA. During this reaction mechanism, cells are repeatedly exposed to reactive oxygen and nitrogen species. Production of FRs in excess amount starts chain reactions that ultimately result in harmful effects. The body of living organisms continuously remains in contact with reactive oxygen species (ROS)/free radicals/superoxide anion. Tsukahara (2007) reported that during normal physiological processes, the balance between FRs and ROS is maintained by antioxidants and enzymatic action. The damaging effect of OS may vary with cellular type and stress imposed during the physiological process. Production of excess FRs affects the functionality of cells either directly or indirectly.

7.3 BIOACTIVE COMPOUNDS IN MILLETS

Documented reports on millets indicate that they are good source of health-beneficial antioxidant compounds. Millets have the potential to combat OS due to the presence of antioxidant properties in them (Kumar et al., 2018; Shahidi & Naczk, 1995; Skrovankova et al., 2015). The health beneficial role and functioning of antioxidants may vary with dose (Rein et al., 2013; Bhat et al., 2015). Millets attract considerable attention from food scientists/dietitians/researchers due to having unique mixture of bioactive compounds with antioxidant potential. Millets rank fourth among major cereals due to the presence of specific nutrients and health benefits.

Millets are a staple food material with their low price and availability. Food industries prefer to use millets as an important substrate for preparing various food products from them. Specificity of nutrient components in millets ensure the consumer satisfaction level along with cost-effective nature. Nowadays, a significant portion of the population is replacing wheat flour with millets due to their gluten-free nature. Inclusion of millet-based products in the diet helps to gain immunity against certain chronic disorders, necessary to sustain routine activity of biological system. Millets' profile indicates that the presence of antioxidants in them ensures significant reduction in the activity of FRs and OS (Zieliński & Kozłowska, 2000). Millets are important industrial crops that have remained unexplored for a long time and need to be studied in detail for their health-promoting features (Gong et al., 2018).

Identified bioactive compounds in millets are ferulic acid, vanillic acid, benzoic acid, cinnamic acid, catechol, ascorbic acid, gallic acid, kaempferol, salicylic acid, syringic acid, protocatechuic acid, and chlorogenic and sinapic acids (Hithamani & Srinivasan, 2014; Chandrasekara et al., 2012; Pradeep & Guha, 2011). Hithamani and Srinivasan (2014) studied finger millet extracts and found gallic acid (3.91 µg/g), *p*-hydroxybenzoic acid (4.41 µg/g), gentisic acid (4.96 µg/g), caffeic acid (5.87 µg/g), syringic acid (60 µg/g), *p*-coumaric acid (1.81 µg/g), sinapic acid (11 µg/g), salicylic acid (5.12 µg/g), and transcinnamic acid (1.55 µg/g). Pearl millet extracts showed the presence of gallic acid (15.3 µg/g), syringic acid (7.4 µg/g), *p*-coumaric acid (1350 µg/g), and ferulic acid (199 µg/g) (Nani et al., 2015). Further, cv. PUSA-415 (pearl millet) extracts during HPLC analysis indicated the presence of ascorbic acid (320 µg/g), gallic acid (120 µg/g), and *p*-coumaric acid (160 µg/g) (Salar et al., 2017b). HPLC analysis of foxtail millet and little millet extracts showed the presence of specific compounds like gallic acid (6.91 and 5.82 µg/g), *p*-hydroxybenzoic acid (1.07 and 4.05 µg/g), vanillic acid (2.37 and 35.69 µg/g), caffeic acid (85.22 and 78.83 µg/g), chlorogenic acid (19.91 and 29.53 µg/g), ferulic acid (93.4 and 129.31 µg/g), sinapic acid (85.36 and 63.24 µg/g), and *p*-coumaric acid (3.66 and 2.71 µg/g) in them. Bioactive compounds present in millets are reported in Tables 7.1 and 7.2.

Table 7.1 Amount of Bioactive Compounds Present in Millets

Millets	TPC	TFC	CTC	References
Pearl millet	3.80–34.1 mg GAE/g	–	25–138.4 mg CE/100 g	Salar et al. (2017b), Salar and Purewal (2017)
Finger millet	15.1–52.3 μmol FAE/g	2.5–12.5 μmol CE/g	2.05–2.6 mg/g	Kumari et al. (2016), Panwar et al. (2016)
Foxtail millet	33.17 mg GAE/100 g	1.4–3.5 μmol CE/g	1.2–2.8 mg/100 g	Sharma et al. (2015)
Proso millet	4.3–10.4 μmol FAE/g	1.3–2.1 μmol CE/g	48–178 mg/100 g	Kumari et al. (2016), Lorenz (1983)
Kodo millet	79.01–359.2 mg GAE/100 g			Sharma et al. (2017)
Barnyard millet	80.14 mg FAE/100 g	58.42 mg CE/100 g	3.25–3.96 mg/g	Pradeep and Sreerama (2015)

7.4 EXTRACTION OF BIOACTIVE COMPOUNDS

During the last 20 years, much attention has been given to the parameters responsible for recovery of phenolics from various natural resources. Extraction is considered as an important process as it brings metabolites with specific activities in extraction medium. Documented extraction reports indicate that there is no single method that can be applied on a variety of substrates to recover all bioactive compounds from them. Methods of extraction need to be optimized for every specific natural resource to get a maximal amount of bioactive compounds. To isolate/extract bioactive compounds, solvent and supercritical fluid extraction methods are generally employed (Ignat et al., 2013; Bucic-Kojic et al., 2007; Bleve et al., 2008). Extraction of active metabolites from experimental material needs grinding, oven drying, shade drying, freeze drying, and defatting. Extraction methodologies not only bring secondary metabolites in extraction media, but nonphenolic residues, acids, and protein may also leach out with phenolics, thus necessitating further purification processes

Table 7.2 Specific Compounds Present in Different Millets

Millets	Specific Compounds	References
Finger millet	Apigenin, caffeic acid, catechin, catechin-O-dihexoside, cinnamic acid, epicatechin, epigallocatechin, gallic acid, gallocatechin, gentisic, kempherol, luteolin, myricetin, p-coumaric acid, p-hydroxybenzoic, procyanidin B1, procyanidin B2, procyanidin dimer, protocatechuic acid, protocatechuic aldehyde, quercetin, rutin, sinapic acid, syringic acid, trans–ferulic, vanillic, vitexin	Xiang et al. (2019), Zhang et al. (2017), Shahidi and Chandrasekara (2013), Banerjee et al. (2012), N'Dri et al. (2012), Pradeep and Guha (2011), Chandrasekara and Shahidi (2011), Shobana et al. (2009), Viswanath et al. (2009), Chethan et al. (2008), Rao and Muralikrishna (2001, 2002, 2004), Sartelet et al. (1996), Watanabe (1999)
Pearl millet	p-Hydroxybenzoic, protocatechuic, vanillic, caffeic acid, p-coumaric, cinnamic, sinapic, trans-ferulic, apigenin, ascorbic acid, myricetin	
Foxtail millet	Gallic acid, p-hydroxybenzoic, protocatechuic, syringic acid, gentisic acid, vanillic caffeic acid, p-coumaric acid, sinapic acid, trans-ferulic catechin, quercetin, apigenin, kempherol	
Proso millet	Gallic acid, p-hydroxybenzoic, protocatechuic, syringic acid, gentisic, vanillic, chlorogenic acid, caffeic acid, p-coumaric acid, sinapic acid, ferulic acid, kempherol, apigenin, myricetin	
Kodo millet	Gallic, p-hydroxybenzoic, protocatechuic, syringic acid, vanillic, chlorogenic acid, caffeic acid, p-coumaric acid, sinapic acid, trans-ferulic, cinnamic acid, kempherol, apigenin, vitexin, isovitexin, luteolin, quercetin	
Little millet	Gallic, protocatechuic, syringic acid, gentisic acid, vanillic, caffeic, p-coumaric acid, sinapic acid, trans-ferulic, apigenin	
Barnyard millet	Luteolin, tricin, N-(p-coumaroyl) serotonin	

(Ignat et al., 2011). Solid–liquid extraction ratio may vary with the type and nature of substrates being extracted. Even within the same substrate, significant variation in phenolics may be observed depending on solvent type used, concentration (Purewal et al., 2020), soil profile in which the substrate was grown, type of chemical fertilizer and biofertilizer used, etc. (Kaur & Purewal, 2019; Salar et al., 2015). The major change that occurs during the extraction phenomenon is the mass transport during which secondary metabolites and other components (sugars, acids, and nonphenolic residues) migrate into the surrounding extraction medium. Under the influence of specific extraction conditions, the majority of phenolics may be recovered (Salar et al., 2016). The mitigation rate of bioactive compounds may be enhanced using specific temperature range, duration of time, concentration gradient, rate of flow, particle size of sample, and other conditions (shaking/static) (Purewal et al., 2020; Corrales et al., 2009) (Figure 7.1). Efficacy of the extraction process is a reflection of process conditions. Extraction parameters have their own effect on overall output (amount of phenolic recovered). Nature of experimental sample also decides the extraction conditions to be used during the process as some resources need harsher conditions/high temperature range and more

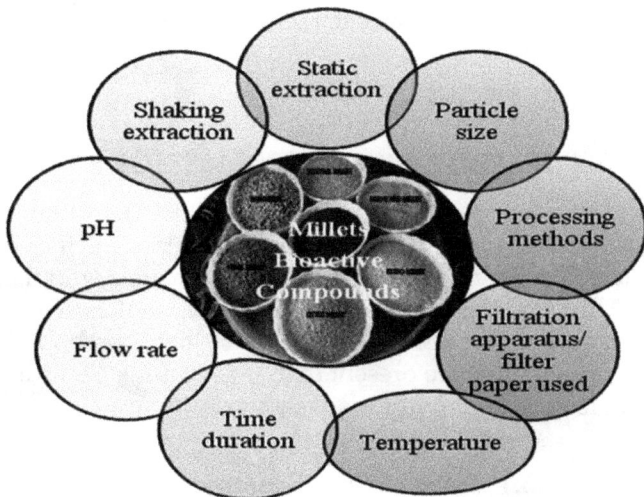

Figure 7.1 Extraction parameters affecting bioactive compounds of millets.

time to release their secondary metabolites; however, in other case moderate conditions are sufficient enough to recover maximum metabolites. Commonly employed liquid extraction media are ethanol (50%); methanol, ethanol, acidified methanol; acetone; water; chloroform; etc. Some biologically active compounds remain in bounds form; these extraction conditions then are not enough as they require specific processing methods to release them in free forms (Salar et al., 2012). Bioactive compounds extracted from millets along with their amount extracted, solvent type, and concentration are reported in Table 7.3.

7.4.1 Quantification of Bioactive Compounds

Scholars/scientists/food industries have their focus on using rapid, sensitive, and selective methods for the evaluation of bioactive profile of extracts prepared from millets and other natural resources. Ample spectroscopic methods have been designed to correctly measure the bioactive compounds present in the extracts. Generally, the methods employed for the detection purpose are either color-based or absorbance-based. Folin–Ciocalteu (FC)/phenol reagent is widely used for the estimation of phenolics in extracts. During the TPC assay, FC reagent and sodium carbonate are used for quantification purpose; however, time duration and concentration of sodium carbonate may vary with the type of modifications and the methods used.

7.5 IDENTIFICATION METHODS

Qualitative and quantification techniques for the detection of specific compounds in extracts are one of the challenging tasks as the compounds present in natural extracts are a mixture of polyphenols with their specific properties. This complex nature of compounds makes it necessary to develop protocols with specific conditions so that compounds may be detected even when they are present in lower amounts. Few identification methods for bioactive compounds are shown in Figure 7.2.

7.5.1 Thin Layer Chromatography

Thin layer chromatography (TLC) is a technique that is generally used to separate out a mixture of volatile compounds. To perform the TLC

Table 7.3 Extraction Solvent and Conc. for the Recovery of Bioactive Compounds from Millets

Millets	Solvent Type	Solvent Conc.	References
Pearl millet	Ethanol	50%	Purewal et al. (2019), Salar et al. (2017b), Salar and Purewal (2017, 2016)
Pearl millet	Water:ethanol:methanol	73%:16%:11%	Purewal et al. (2020)
Pearl millet	Methanol	50%	Salar et al. (2016)
	HCl: methanol	1:99; 0.5:99.5	
Finger millet	Acidified methanol	–	Hithamani and Srinivasan (2014)
Finger millet	Methanol	80%	Xiang et al. (2019)
Foxtail millet Proso millet Finger millet Kodo millet Little millet	Acetone	70%	Chandrasekara et al. (2012)
Kodo millet	Water, methanol, acetone	–	Sharma et al. (2017a)
Foxtail millet Little millet	Aqueous methanol with HCl	1% HCl	Pradeep and Sreerama (2017)
Pearl millet	Petroleum ether, ethyl acetate, ethanol	–	Marmouzi et al. (2018)
Barnyard Foxtail Proso	1% HCl–methanol	–	Pradeep and Sreerama (2015)
Barnyard Italian millet Finger millet	Ethanol	70%	Ofosu et al. (2020)
Foxtail millet	Ethanol:water:acetic acid	70:29.5:0.5	Zhang et al. (2017)
Pearl millet Finger millet	Methanol/formic acid	50%/1%	Hassan et al. (2020)
Foxtail millet	Ethanol	80%	Sharma et al. (2015)
Finger millet	Acetone	70%	Kumari et al. (2020)

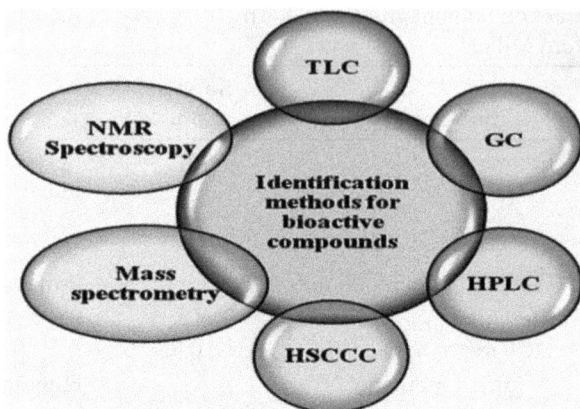

Figure 7.2   Different identification methods for bioactive compounds.

experimentation, the necessary things are glass/plastic sheet, aluminum foil, and coating of suitable absorbent material (silica/cellulose/alumina). Absorbent layer is well known as stationary phase. As soon as the sample is mounted on the suitable place in the plate, absolute solvent/mixture of solvent (mobile phase) is drawn up due to capillary action. Separation of compounds may be achieved as particular analytes ascend the plate at different rates. Difference between the mobile and stationary phases may also be the reason for the separation. Different dark spots may be observed on the TLC plate which could be easily visualized under UV light exposure.

7.5.2 Gas Chromatography

Gas chromatography (GC) could be a good option to identify compounds as it is basically designed to separate out volatile compounds from specific mixtures. The GC method uses vaporization temperature for compound separation through a heated column with their partition between inert gas under pressure and nonvolatile phase which is coated on column (Ignat et al., 2013; Balas & Popa, 2007). Retention time describes the affinity of a particular compound/derivative towards the stationary phase. GC equipment has a hot injection port whose function is to vaporize volatile compounds present in extract sample and to keep them in gaseous form as they cross through the column (Maul et al., 2008; Li et al., 2009;

Fiamegos et al., 2004). However, GC analysis is restricted for the detection of certain specific volatile compounds only.

7.5.3 High-Performance Liquid Chromatography

Among different methods that are capable of identifying bioactive compounds, HPLC is one of the best methods for to analyze the bioactive compounds qualitatively and quantitatively. HPLC analysis usually includes a reversed-phase C18 column, UV–Vis diode array detector, and a binary solvent system containing acidified water (solvent A) and a polar organic solvent (solvent B) (Gini & Jothi, 2018; Krstonosic et al., 2019; Bubenchikov & Goncharov, 2005). Reverse-phase HPLC (RP-HPLC) is widely used as an analysis tool for the detection of bioactive compounds in extracts prepared from natural resources.

7.5.4 High-Speed Counter-Current Chromatography (HSCCC)

Separation of bioactive compounds from extract prepared from natural resources is one of the key requirements of analytical experimentation. The factors that should be of utmost importance during analysis are convenience, less sample loss, high accuracy, maximum resolution, and specific compound recovery without impurities. HSCCC is one of the reliable and sensitive ways of identification in which the stationary phase remains in liquid form and provides good chromatographic details. Further, the stationary phase remains in the column due to centrifugal partition technique, whereas the liquid phase passed with maximum yield and level of purity. Compared with traditional liquid chromatography where only one phase is used to elute the analytes, in HSCCC, two-phase solvent system of immiscible solvents is used where one behaves as the stationary and the other as the mobile phase (Sethi et al., 2009; Gu et al., 2007). The factors that affect the overall success of identification method are (1) sample solubility in solvent, (2) sample stability, (3) settling time, (4) partition coefficient, and (5) stationary-phase retention.

7.5.5 Mass Spectrometry

To evaluate molecular weights and patterns of fragmentation, the commonly used method is mass spectrometry, which is capable enough to identify phenolic compounds. The reliability and sensitivity of mass spectrometry make it unique among other identification methods. The exact molecular weight of compounds present in the sample extract could be evaluated by coupling mass spectrometry with chromatography (Fulcrand et al., 2008; Reed et al., 2005; Alonso-Salcesa et al., 2004).

7.5.6 NMR Spectroscopy

One important and promising method for the identification of complex polyphenols is NMR spectroscopy. The major advantage of using NMR for bioactive compound identification is the method's simplicity, stability, and ease of spectra interpretation. Solid samples from natural resources can be extracted using deuterated solvent, whereas liquid/freeze-dried sample extracts may require addition of deuterated solvent D_2O (5%–10%), which provides a signal for stabilization of the magnetic field and allows optimized NMR peaks with high resolution (Ignat et al., 2011, 2013; Naczk & Shahidi, 2004).

7.6 EFFECT OF PROCESSING ON MILLET PHENOLICS

Millets could be processed using different methods, i.e., (1) solid-state fermentation, (2) steeping/malting, and(3) thermal processing. Different processing methods have significant effects on the bioactive profile of millets; some have positive effects, whereas other methods have negative effects on specific bioactive compounds (Kaur et al., 2019; Salar et al., 2017b; Adebiyi et al., 2017). Processing methods that result in enhancement of antioxidant activity of millets are of utmost importance. Millets are a good source of antioxidants, and processing may modulate the content by a significant amount. Modulation of bioactive profile via solid-state fermentation of natural resources is common as the process is capable of converting bound phenolics to free forms (Li et al., 2020; Maia et al., 2020). Processing helps in achieving the desired characteristics in millet-based food products. One of the points that need to be considered during processing of millets is specific conditions, since, under the effect of specific limits, treatment might result in positive effects; however, beyond that limit, it may deteriorate the quality parameters of products. The effects of different processing methods on phenolic compounds of millets are reported in Table 7.4.

Table 7.4 Effect of Processing Methods on Phenolic Compounds of Millets

Cereal Grains	Processing Type	Effect on TPC	References
Pearl millet	Solid-state fermentation	3.31–33.61 mg GAE/g	Salar and Purewal (2016), Purewal et al. (2019, 2020)
Pearl millet	Thermal processing	6.4–5.4 mg GAE/g	Salar et al. (2017b)
Finger millet	Germination	10.37–9.26 µg GAE/ml	Rani and Antony (2014)
Barnyard millet	Germination	80.14–246.89 mg FAE/100 g	Pradeep and Sreerama (2015)
Foxtail millet	Germination	72.70–170.14 mg FAE/100 g	
Proso millet	Germination	74.43–238.38 mg FAE/100 g	
Barnyard millet	Steaming	80.14–76.08 mg FAE/100 g	
Foxtail millet	Steaming	72.70–76.22 mg FAE/100 g	
Proso millet	Steaming	74.43–79.59 mg FAE/100 g	
Barnyard millet	Microwave	80.14–73.65 mg FAE/100 g	
Foxtail millet	Microwave	72.70–104.46 mg FAE/100 g	
Proso millet	Microwave	74.43–87.43 mg FAE/100 g	
Pearl millet	Malting	764.45–451.92 mg/100 g	Archana and Kawatra (1998)
Pearl millet	Blanching	764.45–544.45 mg/100 g	
Broomcorn millet	Roasting	2.95–6.70 mg FAE/g	Kalam-Azad et al. (2019)
	Steaming	2.95-3.15 mg FAE/g	
	Puffing	2.95–6.45 mg FAE/g	

(Continued)

183

Table 7.4 (*Continued*) Effect of Processing Methods on Phenolic Compounds of Millets

Cereal Grains	Processing Type	Effect on TPC	References
	Extrusion	2.95–4.55 mg FAE/g	
Pearl millet	Roasting	1.70–0.91 mg/g	Obadina et al. (2016)
Finger millet	Roasting	3.14–2.23 mg/g	Singh et al. (2018)
Foxtail millet	Boiling	4.56 mM GAE	Kumar and Vaishnavi (2012)
	Blanching	4.16 mM GAE	
	Roasting	8.36 mM GAE	
	Soaking	9.44 mM GAE	
	Germinating	16.92 mM GAE	
Proso millet	Boiling	1.44 mM GAE	
	Blanching	5.88 mM GAE	
	Roasting	4.36 mM GAE	
	Soaking	5.40 mM GAE	
	Germinating	7.52 mM GAE	

7.6.1 Solid-State Fermentation (SSF)

Scientists are continuously working on different parameters to improve the status of millets at the industrial scale. Among the different methodologies adopted for the millets' nutrient improvement, one of the important ones is SSF. The success of SSF process is governed by several factors that ultimately decide the economic feasibility of the process at the technical level. These factors include substrate selection, starter culture (microbial strains), incubation conditions, and extraction conditions. SSFs of various natural resources for various purposes have been documented, i.e., wheat bran (*A. niger*) (Orzua et al., 2009), maize (*Thamnidium elegans*) (Salar et al., 2012), oats (*A. cinnamomea*) (Yang et al., 2013), rice bran (*Rhizopus oryzae*) (Schmidt et al., 2014), wheat (*A. awamori*) (Sandhu et al., 2016), and maize (*P. ostreatus*) (Acosta-Estrada et al., 2019).

SSF significantly modulates the phenolic compounds of pearl millet by 5.32 fold (6.4–34.1 mg GAE/g dwb) and 3.06 fold (6.11–18.73 mg GAE/g dwb) enhancement in bioactive compounds by *A. sojae* and *A. oryzae* as starter cultures (Salar et al., 2017b; Salar & Purewal, 2016). Adebiyi et al. (2017) observed a significant ($P < 0.05$) effect of SSF on proximate composition of pearl millet in the forms of ash (1.86%–1.36%), crude fiber (1.26%–1.33%), crude protein (5.47%–5.80%), crude fat (2.25%–1.70%), and energy value (355.31–362.54 kcal).

Further enzyme production during the process of SSF useful in converting the bound form of bioactive compounds to their free form. SSF, being a useful method for microbial transformation and modulation of bioactive profiles, is currently being used on different substrates like cereal grains, pulses, and other natural resources (Dhull et al., 2020; Chen et al., 2020; Torres-León et al., 2019; Shin et al., 2019). Substrates (cereal grains, pulses, and industrial wastes) processed through SSF could act as good source of potent antioxidants, important minerals, and amino acids. To eradicate the hunger problem, SSF could prove to be an important approach while preparing functional food products (Postemsky & Curvetto, 2015; Bei et al., 2018; Huang et al., 2017; Zieliński et al., 2019).

7.6.2 Thermal Processing

Roasting is a processing method during which dry heat is used to improve the taste, flavor, and texture of food material. Food material is processed uniformly on all sides at a temperature of 150°C. During roasting, the flavor of food products is enhanced due to Maillard reactions and caramelization on the surface. While roasting, the food product may be placed on a roasting pan or rack to ensure uniform heating. Roasting is adopted by researchers/scientists to modulate the nutritional profile of substrates. Different substrates were evaluated to check the effect of roasting, such as Samh seeds (*Mesembryanthemum forsskalei* Hochst) (Ahmed et al., 2020), buckwheat (Ma et al., 2020), chia seeds (Hatamian et al., 2020), barley (Zhao et al., 2020; Baba et al., 2016; Sharma & Gujral, 2011), chickpea (Jogihalli et al., 2017), brown rice (Shi et al., 2018), cashew nuts (Chandrasekara & Shahidi, 2011), peanut kernels (Bagheri et al., 2016), oats (Gujral et al., 2011), black soybean (Kim et al., 2011), hazelnuts (Özdemir & Devres, 2000), and green gram seeds (Alkaltham et al., 2020).

7.7 CONCLUSION

Documented reports on millets suggest that millet grains and their milling fractions are good source of health-benefiting bioactive compounds with antioxidant potential. Millets could act as essential sources

185

of food and feed, especially in poverty-stricken regions worldwide. For many years, millets' potential for maintaining a disease-free healthy life remained unexplored. Little information and less innovative techniques limited the use of millets at the industrial scale. The nutritional profile of millets demonstrated the presence of specific nutrients and bioactive compounds, which makes them attractive substrates for the preparation of specific products with broad-spectrum health benefits. Knowledge on bioactive compounds of millets can make pharmaceutical industries capable of preparing various disease-controlling millet-based products.

REFERENCES

Acosta-Estrada, B. A., Villela-Castrejón, J., Perez-Carrillo, E., Gómez-Sánchez, C. E. and Gutiérrez-Uribe, J. A. 2019. Effects of solid-state fungi fermentation on phenolic content, antioxidant properties and fiber composition of lime cooked maize by-product (nejayote). *Journal of Cereal Science* 90: 102837.

Adebiyi, J. A., Obadina, A. O., Adebo, O. A. and Kayitesi, E. 2017. Comparison of nutritional quality and sensory acceptability of biscuits obtained from native, fermented and malted pearl millet (*Pennisetum glaucum*) flour. *Food Chemistry* 232: 210–217.

Adom, K. and Liu, R. 2002. Antioxidant activity of grains. *Journal of Agricultural and Food Chemistry* 50: 6182–6187.

Adom, K. K., Sorrells, M. E. and Liu, R. H. 2005. Phytochemicals and antioxidant activity of milled fractions of different wheat varieties. *Journal of Agricultural and Food Chemistry* 53: 2297–2306.

Ahmed, M. I. A., Al-Juhaimi, F. Y., Osman, M. A., Al-Maiman, S. A., Hassan, A. B., Alqah, H. A. S., Babiker, E. E. and Ghafoor, K. 2020. Effect of oven roasting treatment on the antioxidant activity, phenolic compounds, fatty acids, minerals, and protein profile of Samh (*Mesembryanthemum forsskalei* Hochst) seeds. *LWT Food Science and Technology*: 109825.

Alkaltham, M. S., Salamatullah, A. M., Ozcan, M. M., Uslu, N. and Hayat, K. 2020. The effects of germination and heating on bioactive properties, phenolic compounds and mineral contents of green gram seeds. *LWT Food Science and Technology* 134: 110106.

Alonso-Salcesa, R. M., Ndjoko, K., Queiroz, E. F., Ioset, J. R., Hostettmann, K., Berrueta, L. A., Gallo, B. and Vicente, F. 2004. On-line characterisation of apple polyphenols by liquid chromatography coupled with mass spectrometry and ultraviolet absorbance detection. *Journal of Chromatography A* 1046: 89–100.

Anderson, J. W., Baird, P., Davis, R. H., Ferreri, S., Knudtson, M., Koraym, A., Waters, V. and Williams, C. L. 2009. Health benefits of dietary fiber. *Nutrition Reviews* 67: 188–205.

Archana, S. S. and Kawatra, A. 1998. Reduction of polyphenol and phytic acid content of pearl millet grains by malting and blanching. *Plant Foods for Human Nutrition* 53: 93–98.

Baba, W. N., Rashid, I., Shah, A., Ahmad, M., Gani, A., Masoodi, F. A., Wani, I. A. and Wani, S. M. 2016. Effect of microwave roasting on antioxidant and anticancerous activities of barley flour. *Journal of the Saudi Society of Agricultural Sciences* 15: 12–19.

Bagheri, H., Kashaninejad, M., Ziaiifar, A. M. and Aalami, M. 2016. Novel hybridized infrared-hot air method for roasting of peanut kernels. *Innovative Food Science & Emerging Technologies* 37: 106–114.

Balas, A. and Popa, V. I. 2007. On characterization of some bioactive compounds extracted from *Picea abies* bark. *Roumanian Biotechnological Letters* 12: 3209–3215.

Banerjee, S., Sanjay, K., Chethan, S. and Malleshi, N. G. 2012. Finger millet (*Eleusine coracana*) polyphenols: investigation of their antioxidant capacity and antimicrobial activity. *African Journal of Food Science* 6: 362–374.

Bangoura, M. L., Nsor-Atindana, J., Zhu, K., Tolno, M. B., Zhou, H. and Wei, P. 2013. Potential hypoglycaemic effects of insoluble fibres isolated from foxtail millets (*Setaria italica* (L.) P. Beauvois). *International Journal of Food Science and Technology* 48: 496–502.

Bei, Q., Chen, G., Lu, F., Wu, S. and Wu, Z. 2018. Enzymatic action mechanism of phenolic mobilization in oats (*Avena sativa* L.) during solid-state fermentation with *Monascus anka*. *Food Chemistry* 245: 297–304.

Bhat, Z. F., Kumar, S. and Bhat, H. F. 2015. Bioactive peptides of animal origin: a review. *Journal of Food Science and Technology*, 52: 5377–5392.

Bhatti, M. S., Thukral, A. K., Reddy, A. S. and Kalia, R. K. 2016. RSM and ANN-GA experimental design optimization for electrocoagulation removal of chromium. *Trends in Asian Water Environmental Science and Technology*: 3–21.

Bleve, M., Ciurlia, L., Erroi, E., Lionetto, G., Longo, L., Rescio, L., Schettino, T. and Vasapollo, G. 2008. An innovative method for the purification of anthocyanins from grape skin extracts by using liquid and subcritical carbon dioxide. *Separation and Purification Technology* 64: 192–197.

Brasil, T. A., Capitani, C. D., Takeuchi, K. P. and Ferreira, T. A. P. de C. 2015. Physical, chemical and sensory properties of gluten-free kibbeh formulated with millet flour (*Pennisetum glaucum* (L.) R. Br.). *Food Science and Technology (Campinas)* 35: 361–367.

Bubenchikov, R. A. and Goncharov, N. F. 2005. HPLC analysis of phenolic compounds in field violet. *Pharmaceutical Chemistry Journal* 39: 143–144.

Bucic-Kojic, A., Planinic, M., Tomas, S., Bilic, M. and Velic, D. 2007. Study of solid–liquid extraction kinetics of total polyphenols from grape seeds. *Journal of Food Engineering* 81: 236–242.

Canadas, R., Gonzalez-Miquel, M., Gonzalez, E. J., Diaz, I. and Rodriguez, M. 2020. Hydrophobic eutectic solvents for extraction of natural phenolic antioxidants from winery waste water. Separation and Purification Technology: 117590.

Chandrasekara, A., Naczk, M. and Shahidi, F. 2012. Effect of processing on the antioxidant activity of millet grains. Food Chemistry 133: 1–9.

Chandrasekara, A. and Shahidi, F. 2011a. Anti-proliferative potential and DNA scission inhibitory activity of phenolics from whole millet grains. Journal of Functional Foods 3: 159–170.

Chandrasekara, N. and Shahidi, F. 2011b. Antioxidative potential of cashew phenolics in food and biological model systems as affected by roasting. Food Chemistry 129: 1388–1396.

Chen, G., Liu, Y., Zeng, J., Tian, X., Bei, Q. and Wu, Z. 2020. Enhancing three phenolic fractions of oats (Avena sativa L.) and their antioxidant activities by solid-state fermentation with Monascus anka and Bacillus subtilis. Journal of Cereal Science 93: 102940.

Chethan, S., Dharmesh, S. M. and Malleshi, N. G. 2008. Inhibition of aldose reductase from cataracted eye lenses by finger millet (Eleusine coracana) polyphenols. Bioorganic and Medicinal Chemistry 16: 10085–10090.

Choi, Y. Y., Osada, K., Ito, Y., Nagasawa, T., Choi, M. R. and Nishizawa, N. 2005. Effects of dietary protein of Korean foxtail millet on plasma adiponectin, HDL-cholesterol and insulin levels in genetically type 2 diabetic mice. Bioscience, Biotechnology and Biochemistry 69: 31–37.

Corrales, M., García, A. F., Butz, P. and Tauscher, B. 2009. Extraction of anthocyanins from grape skins assisted by high hydrostatic pressure. Journal of Food Engineering 90: 415–421.

Dai, J. and Mumper, R. J. 2010. Plant phenolics: extraction, analysis and their antioxidant and anticancer properties. Molecules 15: 7313–7352.

Dhull, S. B., Punia, S., Kidwai, M. K., Kaur, M., Chawla, P., Purewal, S. S., Sangwan, M. and Palthania, S. 2020. Solid-state fermentation of lentil (Lens culinaris L.) with Aspergillus awamori: effect on phenolic compounds, mineral content, and their bioavailability. Legume Science 2: 1–12.

Dureja, J. S., Singh, R. and Bhatti, M. S. 2014. Optimizing flank wear and surface roughness during hard turning of AISI D3 steel by Taguchi and RSM methods. Production & Manufacturing Research 2: 767–783.

Dzah, C. S., Duan, Y., Zhang, H., Boateng, N. A. S. and Ma, H. 2020. Ultrasound-induced lipid peroxidation: effects on phenol content and extraction kinetics and antioxidant activity of Tartary buckwheat (Fagopyrum tataricum) water extract. Food Bioscience: 100719.

Fardet, A. 2010. New hypotheses for the health-protective mechanisms of whole-grain cereals: what is beyond fibre? Nutrition Research Reviews 23: 65–134.

Fiamegos, Y. C., Nanos, C. G., Vervoort, J. and Stalikas, C. D. 2004. Analytical procedure for the in-vial derivatization-extraction of phenolic acids and flavonoids in methanolic and aqueous plant extracts followed by gas chromatography with mass-selective detection. Journal of Chromatography A 1041: 11–18.

Finkel, T. 2003. Oxidant signals and oxidative stress. Current Opinion in Cell Biology 15: 247–254.

Fulcrand, H., Mane, C., Preys, S., Mazerolles, G., Bouchut, C., Mazauric, J. P., Souquet, J. M., Meudec, E., Li, Y., Cole, R. B. and Cheynier, V. 2008. Direct mass spectrometry approaches to characterize polyphenols composition of complex samples. *Phytochemistry* 69: 3131–3138.

Gini, T. G. and Jothi, G. J. 2018. Column chromatography and HPLC analysis of phenolic compounds in the fractions of Salvinia molesta Mitchell. Egyptian Journal of Basic and Applied Sciences 5: 197–203.

Gong, L., Cao, W., Chi, H., Wang, J., Zhang, H., Liu, J. and Sun, B. 2018. Whole cereal grains and potential health effects: involvement of the gut microbiota. Food Research International 103: 84–102.

Gu, M., Wang, X., Su, Z. and Ouyang, F. 2007. One-step separation and purification of 3,4-dihydroxyphenyllactic acid, salvianolic acid B and protocatechualdehyde from Salvia miltiorrhiza, bunge by high-speed counter-current chromatography. Journal of Chromatography A 1140: 107–111.

Gujral, H. S., Sharma, P. and Rachna, S. 2011. Effect of sand roasting on beta glucan extractability, physicochemical and antioxidant properties of oats. LWT - Food Science and Technology 44: 2223–2230.

Hassan, Z. M., Sebola, N. A. and Mabelebele, M. 2020. Assessment of the phenolic compounds of pearl and finger millets obtained from South Africa and Zimbabwe. Food Science & Nutrition: 1–9.

Hatamian, M., Noshad, M., Abdanan-Mehdizadeh, S. and Barzegar, H. 2020. Effect of roasting treatment on functional and antioxidant properties of chia seed flours. NFS Journal 21: 1–8.

Hithamani, G. and Srinivasan, K. 2014. Effect of domestic processing on the polyphenol content and bioaccessibility in finger millet (*Eleusine coracana*) and pearl millet (*Pennisetum glaucum*). *Food Chemistry* 164: 55–62.

Huang, S., Ma, Y., Zhang, C., Cai, S. and Pang, M. 2017. Bioaccessibility and antioxidant activity of phenolics in native and fermented Prinsepia utilis Royle seed during a simulated gastrointestinal digestion in vitro. *Journal of Functional Foods* 37: 354–362.

Ignat, I., Volf, I. and Popa, V. I. 2011. A critical review of methods for characterisation of polyphenolic compounds in fruits and vegetables. *Food Chemistry* 126: 1821–1835.

Ignat, I., Volf, I., Popa, V. I. 2013. Analytical methods of phenolic compounds. *Natural Products*: 2061–2092.

189

Irakli, M. N., Samanidou, V. F., Biliaderis, C. G. and Papadoyannis, I. N. 2012. Development and validation of an HPLC-method for determination of free and bound phenolic acids in cereals after solid-phase extraction. *Food Chemistry* 134: 1624–1632.

Jogihalli, P., Singh, L., Kumar, K. and Sharanagat, V. S. 2017. Novel continuous roasting of chickpea (*Cicer arietinum*): study on physico-functional, antioxidant and roasting characteristics. *LWT Food Science and Technology* 86: 456–464.

Ju-Sung, K., Tae Kyung, H. and Myong-Jo, K. 2011. The inhibitory effects of ethanol extracts from sorghum, foxtail millet and proso millet on alpha-glucosidase and alpha-amylase activities. *Food Chemistry* 124: 1647–1651.

Kaith, B. S., Saruchi, Jindal, R. and Bhatti, M. S. 2012. Screening and RSM optimization for synthesis of a gum tragacanth–acrylic acid based device for in situ controlled cetirizine dihydrochloride release. *Soft Matter* 8: 2286.

Kalam-Azad, M. O., Jeong, D. I., Adnan, M., Salitxay, T., Heo, J. W., Naznin, M. T., Lim, J. D., Cho, D. H., Park, B. J. and Park, C. H. 2019. Effect of different processing methods on the accumulation of the phenolic compounds and antioxidant profile of broomcorn millet (*Panicum miliaceum* L.) Flour. *Foods* 8: 1–13.

Kaur, P., Dhull, S. B., Sandhu, K. S., Salar, R. K. and Purewal, S. S. 2018. Tulsi (*Ocimum tenuiflorum*) seeds: in vitro DNA damage protection, bioactive compounds and antioxidant potential. *Journal of Food Measurement and Characterization* 12: 1530–1538.

Kaur, P. and Purewal, S. S. 2019. Biofertilizers and their role in sustainable agriculture. In: Giri, B., Prasad, R., Wu, Q. S. and Varma A., editors. *Biofertilizers for sustainable agriculture and environment.* Soil Biology, 55. Cham: Springer.

Kaur, P., Purewal, S. S., Sandhu, K. S., Kaur, M. and Salar, R. K. 2019. Millets: a cereal grain with potent antioxidants and health benefits. *Journal of Food Measurement and Characterization* 13: 793–806.

Kim, H. G., Kim, G. W., Oh, H., Yoo, S. Y., Kim, Y. O. and Oh, M. S. 2011. Influence of roasting on the antioxidant activity of small black soybean (Glycine max L. Merrill). *LWT Food Science and Technology* 44: 992–998.

Krstonosic, M. A., Hogervorst, J. C., Mikulic, M. and Gojković-Bukarica, L. 2019. Development of HPLC method for determination of phenolic compounds on a core shell column by direct injection of wine samples. *Acta Chromatographica* 32: 134–138.

Kumar, A., Tomer, V., Kaur, A., Kumar, V. and Gupta, K. 2018. Millets: a solution to agrarian and nutritional challenges. *Agriculture & Food Security* 7: 31.

Kumar, J. B. M. and Vaishnavi, I. R. 2012. Nutrient and antioxidant analysis of raw and processed minor millets. *Elixir Food Science* 52: 11279–11282.

Kumari, D., Chandrasekara, A., Athukorale, P. and Shahidi, F. 2020. Finger millet porridges subjected to different processing conditions showed low glycemic index and variable efficacy on plasma antioxidant capacity of healthy adults. *Food Production, Processing and Nutrition* 2: 13.

Kumari, D., Madhujith, T. and Chandrasekara, A. 2016. Comparison of phenolic content and antioxidant activities of millet varieties grown in different locations in Sri Lanka. *Food Science & Nutrition* 5: 474–485.

Lee, S. H., Chung, I. M., Cha, Y. S. and Park, Y. 2010. Millet consumption decreased serum concentration of triglyceride and C-reactive protein but not oxidative status in hyperlipidemic rats. Nutrition Research 30: 290–296.

Li, S., Jin, Z., Hu, D., Yang, W., Yan, Y., Nie, X., Lin, J., Zhang, Q., Gai, D., Ji, Y. and Chen, X. 2020. Effect of solid-state fermentation with Lactobacillus casei on the nutritional value, isoflavones, phenolic acids and antioxidant activity of whole soybean flour. LWT Food Science & Technology 125: 109264.

Li, F., Liu, Q., Cai, W. and Shao, X. 2009. Analysis of scopoletin and caffeic acid in tobacco by GC–MS after a rapid derivatization procedure. Chromatographia 69: 743–748.

Lorenz, K. 1983. Tannins and phytate content in proso millets (Pennisetum miliaceum). Cereal Chemistry 60: 424–426.

Ma, Q., Zhao, Y., Wang, H. L., Li, J., Yang, Q.-H., Gao, L. C., Murat, T. and Feng, B. L. 2020. Comparative study on the effects of buckwheat by roasting: antioxidant properties, nutrients, pasting, and thermal properties. Journal of Cereal Science: 103041.

Maia, D. C. I., Thomaz dos Santos D'Almeida, C., Guimarães Freire, D. M., Cavalcanti, E., d'Avila, C., Cameron, L. C., Dias, J. F. and Larraz Ferreira, M. S. 2020. Effect of solid-state fermentation over the release of phenolic compounds from brewer's spent grain revealed by UPLC-MSE. LWT Food Science and Technology 133: 110136.

Marmouzi, I., Ali, K., Harhar, H., Gharby, S., Sayah, K., et al. 2018. Functional composition, antibacterial and antioxidative properties of oil and phenolics from Moroccan Pennisetum glaucum seeds. Journal of the Saudi Society of Agricultural Sciences 17: 229–234.

Maul, R., Schebb, N. H. and Kulling, S. E. 2008. Application of LC and GC hyphenated with mass spectrometry as tool for characterization of unknown derivatives of isoflavonoids. *Analytical Bioanalytical Chemistry* 391: 239–250.

Naczk, M. and Shahidi, F. 2004. Extraction and analysis of phenolics in food. *Journal of Chromatography A* 1054: 95–111.

Nani, A., Belarbi, M., Ksouri-Megdiche, W., Abdoul-Azize, S., Benammar, C., Ghiringhelli, F., Hichami, A. and Khan, N. A. 2015. Effects of polyphenols and lipids from *Pennisetum glaucum* grains on T-cell activation: modulation of Ca^{2+} and ERK1/ERK2 signaling. *BMC Complementary Alternative Medicine* 15: 426.

N'Dri, D., Mazzeo, T., Zaupa, M., Ferracane, R., Fogliano, V. and Pellegrini, N. 2012. Effect of cooking on the total antioxidant capacity and phenolic profile of som e whole-meal African cereals. *Journal of the Science of Food and Agriculture* 93: 29–36.

Niro, S., D'Agostino, A., Fratianni, A., Cinquanta, L. and Panfili, G. 2019. Gluten-free alternative grains: nutritional evaluation and bioactive compounds. *Foods* 8: 208.

Obadina, A., Ishola, I. O., Adekoya, I. O., Soares, A. G., de Carvalho, C. W. P. and Barboza, H. T. 2016. Nutritional and physico-chemical properties of flour from native and roasted whole grain pearl millet *(Pennisetum glaucum* [L.]R. Br.). *Journal of Cereal Science* 70: 247–252.

Ofosu, F. K., Elahi, F., Daliri, E. B. M., Chelliah, R., Ham, H. J., Kim, J. H., Han, S. N., Hur, J. H. and Oh, D. H. 2020. Phenolic profile, antioxidant, and antidiabetic potential exerted by millet grain varieties. *Antioxidants* 9: 254.

Orzua, M. C., Mussatto, S. I., Contreras-Esquivel, J. C., Rodriguez, R., la Garza, H., Teixeira, J. A. and Aguilar, C. N. 2009. Exploitation of agro industrial wastes as immobilization carrier for solid-state fermentation. *Industrial Crops and Products* 30: 24–27.

Özdemir, M. and Devres, O. 2000. Analysis of color development during roasting of hazelnuts using response surface methodology. *Journal of Food Engineering* 45: 17–24.

Panwar, P., Dubey, A. and Verma, A. K. 2016. Evaluation of nutraceutical and antinutritional properties in barnyard and finger millet varieties grown in Himalayan region. *Journal of Food Science and Technology* 53: 2779–2787.

Postemsky, P. D. and Curvetto, N. R. 2015. Solid-state fermentation of cereal grains and sunflower seed hulls by *Grifola gargal* and *Grifola sordulenta*. *International Biodeterioration & Biodegradation* 100: 52–61.

Pradeep, S. R. and Guha, M. 2011. Effect of processing on the nutr aceutical and antioxidant properties of little millet (Pannicum sumatrense) extracts. Food Chemistry 126: 1643–1647.

Pradeep, P. M. and Sreerama, Y. N. 2015. Impact of processing on the phenolic profiles of small millets: evaluation of their antioxidant and enzyme inhibitory properties associated with hyperglycemia. Food Chemistry 169: 455–463.

Pradeep, P. M. and Sreerama, Y. N. 2017. Soluble and bound phenolics of two different millet genera and their milled fractions: comparative evaluation of antioxidant properties and inhibitory effects on starch hydrolysing enzyme activities. Journal of Functional Foods 35: 682–693.

Purewal, S., Salar, R., Bhatti, M., Sandhu, K. S., Singh, S. K. and Kaur, P. 2020. Solid-state fermentation of pearl millet with Aspergillus oryzae and Rhizopus azygosporus: effects on bioactive profile and DNA damage protection activity. *Journal of Food Measurement and Characterization* 14: 150–162.

Purewal, S. S., Sandhu, K. S., Salar, R. K. and Kaur, P. 2019. Fermented pearl millet: a product with enhanced bioactive compounds and DNA damage protection activity. Journal of Food Measurement and Characterization 13: 1479–1488.

Rani, R. S. and Antony, U. 2014. Effect of germination and fermentation on polyphenols in finger millet (*Eleusine coracana*). *International Journal of Food and Nutritional Sciences* 3: 65–68.

Rao, M. V. S. S. T. S. and Muralikrishna, G. 2001. Non-starch polysaccharides and bound phenolic acids from native and malted finger millet (Ragi, *Eleusine coracana*, Indaf-15). *Food Chemistry* 72: 187–192.

Rao, M. V. S. S. T. S. and Muralikrishna, G. 2002. Evaluation of the antioxidant properties of free and bound phenolic acids from native and malted finger millet (Ragi, Elucine coracana In daf 15). *Journal of Agricultural and Food Chemistry* 50: 889–892.

Rao, M. V. S. S. T. S. and Muralikrishna, G. 2004. Structural analysis of arabinoxylans isolated from native and malted finger millet (*Eleusine coracana*, ragi). *Carbohydrate Research* 339: 2457–2463.

Reed, J. D., Krueger, C. G. and Vestling, M. M. 2005. MALDI-TOF mass spectrometry of oligomeric food polyphenols. *Phytochemistry* 66: 2248–2263.

Rein, M. J., Renouf, M., Cruz-Hernandez, C., Actis-Goretta, L., Thakkar, S. K. and da Silva Pinto, M. 2013. Bioavailability of bioactive food compounds: a challenging journey to bioefficacy. *British Journal of Clinical Pharmacology* 75: 588–602.

Salar, R. K., Certik, M. and Brezova, V. 2012. Modulation of phenolic content and antioxidant activity of maize by solid state fermentation with *Thamnidium elegans* CCF 1456. *Biotechnology and Bioprocess Engineering* 17: 109–116.

Salar, R. K. and Purewal, S. S. 2016. Improvement of DNA damage protection and antioxidant activity of biotransformed pearl millet (*Pennisetum glaucum*) cultivar PUSA-415 using *Aspergillus oryzae* MTCC 3107. *Biocatalysis and Agricultural Biotechnology* 8: 221–227.

Salar, R. K. and Purewal, S. S. 2017. Phenolic content, antioxidant potential and DNA damage protection of pearl millet (*Pennisetum glaucum*) cultivars of North Indian region. *Journal of Food Measurement and Characterization* 11: 126–133.

Salar, R. K., Purewal, S. S. and Bhatti, M. S. 2016. Optimization of extraction conditions and enhancement of phenolic content and antioxidant activity of pearl millet fermented with *Aspergillus awamori* MTCC-548. *Resource Efficient Technologies* 2: 148–157.

Salar, R. K., Purewal, S. S. and Sandhu, K. S. 2017a. Relationships between DNA damage protection activity, total phenolic content, condensed tannin content and antioxidant potential among Indian barley cultivars. *Biocatalysis and Agricultural Biotechnology* 11: 201–206.

Salar, R. K., Purewal, S. S. and Sandhu, K. S. 2017b. Fermented pearl millet (*Pennisetum glaucum*) with in vitro DNA damage protection activity, bioactive compounds and antioxidant potential. *Food Research International* 100: 204–210.

Salar, R. K., Sharma, P. and Purewal, S. S. 2015. In vitro antioxidant and free radical scavenging activities of stem extract of *Euphorbia trigona* Miller. *TANG [Humanitas Medicine]* 5: 1–6.

Sandhu, K. S., Kaur, P., Siroha, A. K. and Purewal, S. S. 2020. Phytochemicals and antioxidant properties in pearl millet. In: Punia, S., Siroha, A. K., Sandhu, K. S., Gahlawat, S. K. and Kaur, M., editors. *Pearl millet: properties, functionality and its applications.* Boca Raton, London, New York: CRC Press. pp. 33–50.

Sandhu, K. S. and Punia, S. 2017. Enhancement of bioactive compounds in barley cultivars by solid substrate fermentation. *Journal of Food Measurement and Characterization* 11: 1355–1361.

Sandhu, K. S., Punia, S. and Kaur, M. 2016. Effect of duration of solid state fermentation by *Aspergillus awamorinakazawa* on antioxidant properties of wheat cultivars. *LWT Food Science and Technology* 71: 323–328.

Sartelet, H., Serghart, S., Lobstain, A., Ingenbleek, Y., Anton, R., Petitfrere, E., Aguie-Aguie, G., Martiny, L. and Haye, B. 1996. Flavonoids extracted from Fonio millet (*Digitaria exilis*) reveal potent anti-thyroid properties. *Nutrition* 12: 100–106.

Schmidt, C. G., Gonçalves, L. M., Prietto, L., Hackbart, H. S. and Furlong, E. B. 2014. Antioxidant activity and enzyme inhibition of phenolic acids from fermented rice bran with fungus *Rhizopus oryzae*. *Food Chemistry* 146: 371–377.

Sethi, N., Anand, A., Sharma, A., Chandrul, K. K., Jain, G. and Srinivasa, K. S. 2009. High speed counter current chromatography: a support-free LC technique. *Journal of Pharmacy and Bioallied Sciences* 1: 8–15.

Shahidi, F. and Chandrasekara, A. 2013. Millet grain phenolics and their role in disease risk reduction and health promotion: a review. *Journal of Functional Foods* 5: 570–581.

Shahidi, F. and Naczk, M. 1995. *Food phenolics: sources, chemistry, effects and applications.* Lancaster: Technomic Publishing Co.

Sharma, P. and Gujral, H. S. 2011. Effect of sand roasting and microwave cooking on antioxidant activity of barley. *Food Research International* 44: 235–240.

Sharma, J., Kaith, B. S. and Bhatti, M. S. 2018. Fabrication of biodegradable superabsorbent using RSM design for controlled release of KNO_3. *Journal of Polymers and the Environment* 26(2): 518–531.

Sharma, N. and Niranjan, K. 2017. Foxtail millet: properties, processing, health benefits, and uses. *Food Reviews International* 34: 329–363.

Sharma, S., Saxena, D. C. and Riar, C. S. 2015. Antioxidant activity, total phenolics, flavonoids and anti-nutritional characteristics of germinated foxtail millet (*Setaria italica*). *Cogent Food & Agriculture* 1: 1081728.

Sharma, S., Sharma, N., Handa, S. and Pathania, S. 2017a. Evaluation of health potential of nutritionally enriched Kodo millet (*Eleusine coracana*) grown in Himachal Pradesh, India. *Food Chemistry* 214: 162–168.

Sharma, J., Sukriti, Kaith, B. S. and Bhatti, M. S. 2017b. Fabrication of biodegradable superabsorbent using RSM design for controlled release of KNO_3. *Journal of Polymers and the Environment* 26: 518–531.

Shi, Y., Wang, L., Fang, Y., Wang, H., Tao, H., Pei, F., Li, P., Xu, B. and Hu, Q. 2018. A comprehensive analysis of aroma compounds and microstructure changes in brown rice during roasting process. *LWT Food Science and Technology* 98: 613–621.

Shin, H. Y., Kim, S. M., Lee, J. H. and Lim, S. T. 2019. Solid-state fermentation of black rice bran with *Aspergillus awamori* and *Aspergillus oryzae*: effects on phenolic acid composition and antioxidant activity of bran extracts. *Food Chemistry* 272: 235–241.

Shobana, S., Sreerama, Y. N. and Malleshi, N. G. 2009. Composition and enzyme inhibitory properties of finger millet (*Eleusine coracana* L.) seed coat phenolics: mode of inhibition of a glucosidase and pancreatic amylase. *Food Chemistry* 115: 1268–1273.

Singh, N., David, J., Thompkinson, D. K., Seelam, B. S., Rajput, H. and Morya, S. 2018. Effect of roasting on functional and phytochemical constituents of finger millet (*Eleusine coracana* L.). *The Pharma Innovation Journal* 7: 414–418.

Singh, T., Dureja, J. S., Dogra, M. and Bhatti, M. S. 2019. Multi-response optimization in environment friendly turning of AISI 304 austenitic stainless steel. *Multidiscipline Modeling in Materials and Structures* 15: 538–558.

Sireesha, Y., Kasetti, R. B., Nabi, S. A., Swapna, S. and Apparao, C. 2011. Anti-hyperglycemic and hypolipidemic activities of *Setaria italica* seeds in STZ diabetic rats. *Pathophysiology* 18: 159–164.

Siroha, A. K. and Sandhu, K. S. 2017. Effect of heat processing on the antioxidant properties of pearl millet (*Pennisetum glaucum* L.) cultivars. *Journal of Food Measurement and Characterization* 11: 872–878.

Siroha, A. K., Sandhu, K. S. and Kaur, M. 2016. Physicochemical, functional and antioxidant properties of flour from pearl millet varieties grown in India. *Journal of Food Measurement and Characterization* 10: 311–318.

Skoronski, E., Fernandes, M., Malaret, F. J. and Hallett, J. P. 2020. Use of phosphonium ionic liquids for highly efficient extraction of phenolic compounds from water. *Separation and Purification Technology*: 117069.

Skrovankova, S., Sumczynski, D., Mlcek, J., Jurikova, T. and Sochor, J. 2015. Bioactive compounds and antioxidant activity in different types of berries. *International Journal of Molecular Sciences* 16: 24673–24706.

Taylor, J. R. N. and Emmambux, M. N. 2008. Gluten-free foods and beverages from millets. *Gluten-Free Cereal Products and Beverages*: 119–148 IV–V.

Torres-León, C., Ramírez-Guzmán, N., Ascacio-Valdés, J., Serna-Cock, L., dos Santos Correia, M. T., Contreras-Esquivel, J. C. and Aguilar, C. N. 2019. Solid-state fermentation with *Aspergillus niger* to enhance the phenolic contents and antioxidative activity of Mexican mango seed: a promising source of natural antioxidants. *LWT Food Science and Technology* 112: 108236.

Tsukahara, H. 2007. Biomarkers for oxidative stress: clinical application in pediatric medicine. *Current Medicinal Chemistry* 14: 339–351.

Viswanath, V., Urooj, A. and Malleshi, N. G. 2009. Evaluation of antioxidant and antimicrobial properties of finger millet polyphenols (*Eleusine coracana*). *Food Chemistry* 114: 340–346.

Watanabe, M. 1999. Antioxidative phenolic compounds from Japanese Barnyard Millet (*Echinochloa utilis*) grains. *Journal of Agricultural and Food Chemistry* 47: 4500–4505.

Willcox, J. K., Ash, S. L. and Catignani, G. L. 2004. Antioxidants and prevention of chronic disease. *Critical Reviews in Food Science and Nutrition* 44: 275–295.

Xiang, J., Apea-Bah, F. B., Ndolo, V. U., Katundu, M. C. and Beta, T. 2019. Profile of phenolic compounds and antioxidant activity of finger millet varieties. *Food Chemistry* 275: 361–368.

Yang, F. C., Yang, Y. H. and Lu, H. C. 2013. Enhanced antioxidant and antitumor activities of *Antrodia cinnamomea* cultured with cereal substrates in solid state fermentation. *Biochemical Engineering Journal* 78: 108–113.

Zhang, L., Li, J., Han, F., Ding, Z. and Fan, L. 2017. Effects of different processing methods on the antioxidant activity of 6 cultivars of foxtail millet. *Journal of Food Quality* 2017: 1–9.

Zhao, B., Shang, J., Liu, L., Tong, L., Zhou, X., Wang, S., Zhang, Y., Wang, L. and Zhou, S. 2020. Effect of roasting process on enzymes inactivation and starch properties of highland barley. *International Journal of Biological Macromolecules* 165: 675–682.

Zieliński, H. and Kozłowska, H. 2000. Antioxidant activity and total phenolics in selected cereal grains and their different morphological fractions. *Journal of Agricultural and Food Chemistry* 48: 2008–2016.

Zieliński, H., Szawara-Nowak, D., Bączek, N. and Wronkowska, M. 2019. Effect of liquid-state fermentation on the antioxidant and functional properties of raw and roasted buckwheat flours. *Food Chemistry* 271: 291–297.

Zulak, K., Liscombe, D., Ashihara, H. and Facchini, P. 2006. *Alkaloids. Plant secondary metabolism in diet and human health*. Oxford, UK: Blackwell Publishing. pp. 102–136.

8

Millet-Based Food Products

8.1 INTRODUCTION

Cereal-based foods have been consumed as a staple food worldwide in various forms, i.e., bakery products, noodles and pasta, snack foods, breakfast cereals, and others (Tebben et al., 2018; Xu et al., 2019). Cereal grains have served as an important part of the human diet because of their minerals, vitamins, proteins, and antioxidant nature (Dias-Martins et al., 2018; Purewal et al., 2020). Diverse availability of gluten-free bakery goods increases the food choice for individuals on a gluten-free diet. The interest in the development of quality gluten-free products has increased because of the presence of population groups that must follow a gluten-free diet. This includes celiac sufferers or wheat-allergic individuals and those who have a non-celiac gluten sensitivity (Reilly, 2016).

The results of various studies demonstrated that the major health-benefiting cereal grains to be added in diet are rice, maize, barley, sorghum, wheat, oat, millets, and rye (Mridula & Sharma, 2015; Kaur et al., 2019). Awolu (2017) noted the utilization of wheat grains for flour production (partial utilization/non-utilization) and to mitigate celiac disease, diversify raw materials for nutritionally rich crops, and reduce importing cost. Millets are important members of *Poaceae* family, which are being cultivated on dry lands (marginal) in subtropical and tropical regions. Millet grains are rich in starch, protein, fat, important dietary fibers, minerals, bioactive compounds, and vitamins (Salar & Purewal, 2017; Chandrasekara & Shahidi, 2010). Due to their gluten-free nature and low glycemic index (GI), millet grains have generated considerable interest of consumers (Saleh et al., 2013; Annor et al., 2017). Millets are nowadays considered superior to wheat and rice as the whole grains and their products provide the proteins necessary for body development, antioxidants, minerals, and vitamins (Taylor & Emmanbux, 2008; Purewal et al., 2019). Millets are gluten-free pseudocereals, which makes them beneficial due to their ability to decrease GI of the foods (McSweeney et al., 2017). This chapter will summarize the various value-added products of millets.

8.2 BAKERY PRODUCTS

Baking process is one of the oldest and important methods for processing grains (Kulkarni et al., 2010). Bakery products are widely used as ready-to-serve snack products worldwide. Bakery products include snacks, breakfast item, and other staple products such as bread (Misra & Tiwari, 2014). The popularity of biscuits, bread, pastries, cakes, muffins, etc. signal a new era in the food technology that makes the millet grains important at the industrial level. Different millets products are shown in Figure 8.1.

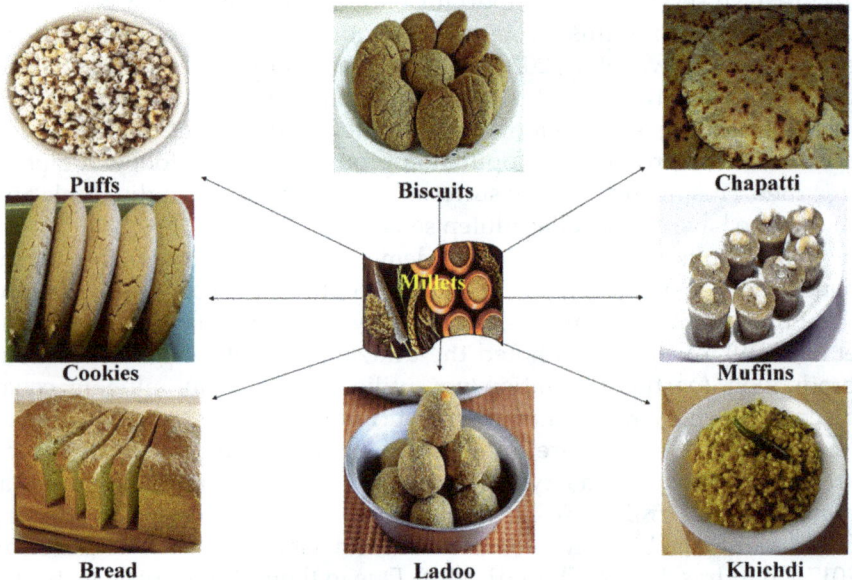

Figure 8.1 Different millet products.

8.2.1 Cookies

Cookies represent the largest category of snacks in the bakery industry and can serve as effective vehicles of nutrient supply to the consumer. The word "cookie" refers to small cakes, derived from the Dutch words *kockje* or *koekie*. They refer to the baked products containing three major ingredients: flour, fat, and sugar. Cookies have low water content (1%–5%) and can also contain minor ingredients like leavening agents, salt, emulsifiers, and yeast (Pareyt & Delcour, 2008). Sharma et al. (2016) reported gluten-free cookies from minor millets (foxtail, barnyard, and kodo millets). The sensory evaluation revealed that cookies prepared from incorporation of germinated foxtail, barnyard, and kodo millets in the proportions of 70:20:10, respectively, are most acceptable, highly nutritious, and have desirable functional properties. Cookies were prepared with maida and pearl millet flour (PMF) by Kulthe et al. (2018). They also observed decrease in carbohydrates and proteins and increase in moisture, fat, ash, and crude fiber with the addition of PMF in maida. The mineral content with respect to calcium, phosphorus, and iron is increased on substitution of maida with PMF, whereas calorific value was found to be slightly decreased. Awolu et al. (2017) prepared cookies from different flour samples (soybean, rice, millets, and tigernut) which were further subjected to different processing protocols such as fermentation, debranning, and malting. Flour blends with fermented millets showed the highest score for overall acceptability. The output of their investigation showed that flour quality and biscuits could be improved using processing techniques and composite flour. Cookies are commonly evaluated in terms of spread factor, crack pattern, breaking strength, and sensory analysis as the most important quality attribute (Xu et al., 2020).

8.2.2 Biscuits

The term "biscuit" is derived from the Latin word *panis biscotis* which means twice-cooked bread. The original process includes baking the biscuits (on hot oven) and then drying (on cool oven) (Misra & Tiwari, 2014). Krishnan et al. (2011) evaluated the biscuits prepared from composite flour of finger millets (seed coat based). Sensory evaluation demonstrated that biscuits prepared from native seed coats (10%) along with hydrothermally processed millets and composite flour of malted millets (20%) could be effective. Saha et al. (2011) have also studied biscuits prepared from finger millets and wheat flour. Hardness of biscuit dough measured by textural profile

199

analysis was more at the 60:40 (finger millet:wheat flour) combination than at the 70:30 one. Studies indicated that the finger millet and wheat flour ratio of 60:40 was better, particularly in the case of biscuit quality. Adebiyi et al. (2017) prepared 100% millet flour-based biscuits by using native, fermented, and malted pearl millet flour. Biscuits prepared with malted and fermented millet flour had more consumer acceptance compared to native flour. Biscuits based on malted millet flour had the best aroma, taste, and overall likeness thanks to their sweeter taste and better flavor, whereas fermented millet biscuits and native millet biscuits had an unpleasant aroma and a relatively bitter taste. Fermentation and malting enhanced the nutritional characteristics of biscuits; they provide improved amino acid profile, mineral bioavailability, and increased phenolic compounds in the treated samples. Singh and Kumar (2018) optimized biscuit formulation made of foxtail millet flour, copra meal flour, and amaranth flour, skim milk powder, and fat and found that relatively higher amounts of amaranth (40%) and fat (42%) resulted in more desired biscuit spread ratio and breaking strength and improved the overall sensory aspect of biscuits.

8.2.3 Muffins

Muffins are considered as high-calorie sweet baked products appreciated by consumers (of all age groups) due to their reliable cost, tasty nature, and texture (Goswami et al., 2015; Mildner-Szkudlarz et al., 2016). Goswami et al. (2015) evaluated muffins prepared from barnyard millet in terms of textural, physical, and sensory properties. The instrumental data indicated that the specific gravity of the batter increased, while weight, baking height, and hardness of muffin samples decreased with increasing proportion of barnyard millet flour (BMF) in flour blends. Muffin samples, including muffins prepared from BMF only, were well accepted by the panelists during sensory evaluation. Rajiv et al. (2011) investigated the effect of replacement of wheat flour with 0%, 20%, 40%, 60%, 80%, and 100% finger millet flour (FMF). The microscopy of muffin batter showed that addition of above 60% FMF in blends decreased the number of air cells, indicating poor air incorporation. With the increase in the FMF level from 0% to 100%, the muffin batter density, viscosity, volume, and total score decreased, whereas crumb firmness increased. Adverse effects on the quality characteristics of cake were observed above 60% FMF. Use of a combination of polysorbate-60 and hydroxyl propyl methylcellulose significantly improved batter characteristics of muffins. The differences between the nutritional values of muffins with and without gluten are less important than with other products, where gluten plays an essential role. However, the gluten-free products present the highest variability in nutritional values, compared to the products with gluten (Belorio & Gómez, 2020).

8.2.4 Breads

Bread is a bakery product whose main ingredients are water, flour, salt, yeast, sugar, and fat which are mixed and fermented to form a viscoelastic dough before being baked (Goesaert et al., 2009). Bread is an important breakfast item that is rich in health-benefiting nutrients. Nutritional quality of bread can be improved using different combinations of whole grains (Onyango et al., 2020). Sarabhai et al. (2020) investigated the effect of enzymes glucose oxidase (GO), xylanase (XYL), and protease (PR) (0.05 and 0.1 g/100g) on textural, rheological, pasting, and sensory qualities of the gluten-free foxtail millet bread. Addition of enzymes significantly increased ($P < 0.05$) specific volume and crumb springiness, while crumb hardness and cohesiveness decreased in comparison to the control. As regards sensory properties, PR-added bread has better taste, aroma, and higher acceptability. It was thus concluded that PR enzyme at 0.1 g/100g can be used to prepare foxtail millet bread with better textural and sensorial properties. Bhol and Bosco (2014) utilized the malted finger millet (MFM) and red kidney bean (RKF) flour for making bread. Bread with MFM showed better sensory score and textural attributes as compared to RKF-incorporated bread. However, the addition of RKF resulted in higher nutritional and mineral composition when compared with the MFM-substituted bread. Finger millet bread is prepared by using hydrothermally treated (HTT) and native finger millet flour (NAT) (Onyango et al., 2020). The higher specific volume and lower crumb firmness and chewiness of WHE-HTT compared to WHE-NAT bread were observed, which may be due to the high α-amylase activity and water absorption capacity of HTT finger millet. Wheat-HTT bread had higher dietary fiber, phytate, and phenolic acid contents but the same starch and protein digestibility as WHE (wheat) bread. Tomić et al. (2020) prepared gluten-free bread by substitution of millet flour with pea, rice, and whey protein concentrate. Millet bread prepared by using whey proteins had the highest specific volume. Overall, the substitution of millet flour by pea, rice, and whey proteins caused a significant reduction of bread hardness and complete loss of bitter taste originating from the millet. The most popular type of flat bread is injera, a flexible, spongy, pancake-like product perforated with "eyes" (Ebba, 1969). Before it is cooked, the soft dough is fermented twice over a period of 2–3 days, acquiring its characteristic sourness and flavor. To cook injera, the dough is thinned to a thick batter and poured onto a lightly oiled pan, which is then covered with a tightly fitting lid to retain the steam (Parker et al., 1989).

8.2.5 Pasta Products

Pasta is a type of food which is prepared from unleavened dough of wheat flour mixed with water or eggs, formed into sheets or other shapes, and consumed after cooking. Gull et al. (2015) studied pasta products prepared from durum wheat semolina along with finger millet flour (FMF), pearl millet flour (PMF), and carrot pomace powder (CPP). It was observed that with increase in the substitution level of millet flours and CPP, solid loss increased and weight gain and firmness decreased. The control pasta showed the lowest solid loss (7.66%) and highest weight gain (33.93 g/10 g) and firmness (5.94 N), while supplemented pasta showed increased solid loss (10%–24.40%) and decreased weight gain (32.3–25.07 g/10 g) and firmness (4.24–2.14 N).

Particle size and blend composition significantly affect the quality of pasta prepared from wheat semolina-pearl millet (Jalgaonkar & Jha, 2016). Increasing concentration of pearl millet flour in blend composition results in an enhancement of ash, protein, and cooking loss of pasta increased with decreasing trend in cohesiveness, hardness, gumminess, springiness and chewiness. A desirable quality is found in pasta prepared from the mixture of wheat semolina and pearl millet flour in the ratio of 70:30. Jalgaonkar et al. (2018) prepared pasta using wheat semolina and pearl millet flour in the ratio of 50:50 fortified with soy flour (DSF), carrot power (CP), mango peel powder (MPP), and moringa leaf power (MLP). Sensory evaluation revealed that maximum incorporation of 15% DSF, 10% CP, 5% MPP, and 3% MLP was found to be suitable in terms of color, cooking loss (< 8%), hardness, and sensory quality. Cordelino et al. (2019) studied various properties of pasta prepared from proso millet. They observed that millet pasta contained less RDS than commercial gluten-free pasta; however, millet and commercial gluten-free pasta had lower protein digestibility than wheat pasta. Sensory panelists detected more graininess and starchiness in millet samples than in commercial pasta. Higher amylose content affects firmness and chewiness of millet pasta.

8.3 FLAT BREADS

In India, an unfermented pancake, called roti, is produced from pearl millet, small millets, sorghum, or maize flour. The results from the study of Panghal et al. (2006) showed wheat flour was less efficient due to the less amount of lysine. Finger millet as compared to wheat possesses significantly higher amounts of lysine, valine, and threonine, which makes it important while balancing the amino acid profile of products (Ravindran, 1992). Siroha et al. (2016) studied chapatti prepared from pearl millet flour and observed its antioxidant properties. It was observed that baking of chapatti decreases the total phenolic content, total flavonoids, and DPPH activity, while the reverse was observed for metal-chelating activity. Panghal et al. (2019) reported that finger millet flour could also be used to prepare chapatti. They observed that with finger millet (flour) addition, puffing height and puffing percentage of chapattis decreased, while baking loss (%) and shrinkage (%) increased. Sharma and Gujral (2019) studied flat bread quality by replacing wheat flour (partly) with minor millets (foxtail, finger, kodo, barnyard, little, and proso) in the proportion of 3:1. Flat breads prepared from wheat millet (composite flour) showed more bake loss and shrinkage with reduced puffing and retrogradation of starch. It is observed that consumer acceptability decreased for wheat–millet composite flour flat breads.

8.4 POPPED/PUFFED MILLETS

Processing of cereal grains through popping/puffing is a traditional practice of grain cooking, and the end product is used as snack/breakfast directly or in combination with some spices/salt/sweeteners (Jaybhaye et al., 2014). Popping is a method of applying dry heat on seed kernels till the expansion of internal moisture; thereafter, the processed material may be used in the preparation of ready-to-eat healthy snacks/weaning foods (Chauhan & Sarita, 2018). The method for popping of cereals is shown in Figure 8.2. Explosion puffing is characterized by sudden release and expansion of water vapor (Sullivan & Craig, 1984). Malleshi and Desikachar (1981) reported that the optimal conditions for ragi puffing were moisture (19%) and equilibration (4 h), followed by sand puffing (270°C). Foxtail millet, barnyard millet, finger millet, and proso millet could be popped with the maximum popping yield (92.77%); the maximum expansion volume (6.51) was reported for proso millet followed by finger millet, foxtail millet, and

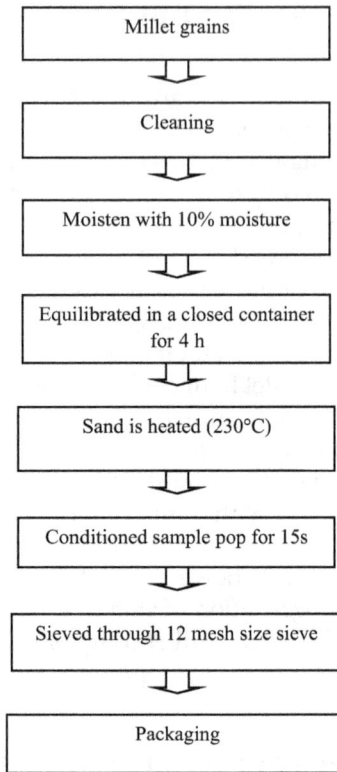

Figure 8.2 Method for popping of cereals (Choudhury et al., 2011).

barnyard millet (Srivastava & Batra, 1998). Choudhury et al. (2011) demonstrated that foxtail millet could be processed using popping. They observed that popping process resulted in significant enhancement in starch digestibility and protein content as compared to untreated native counterparts. Popped samples showed significantly lower values for crude fat and crude fiber as compared to untreated raw millet (yellow and purple varieties), while significantly higher values were observed for carbohydrate and energy. Nutritional and anti-nutritional profile of finger millet was assessed in terms of popping effect (Chauhan & Sarita, 2018). It was found that application of popping process resulted in enhancement of carbohydrate content but with the reverse trend for protein content. A sharp decrease in tannin, oxalic acid, and phytic acid was observed after popping.

8.5 EXTRUDED SNACKS

Snacks are usually prepared from wheat, corn, rice, and oats (Moore, 1994). Extrusion cooking is a cost-effective, versatile, and short-duration process in which moistened, starchy food materials are plasticized and cooked in an enclosed barrel with single or two screws. Kharat et al. (2019) reported that the extrusion process has a series of operations including feed transport, mixing, and forming. Linear programming analysis (LPA) is an important research technique that is being used to model complex multifactorial problems (MFP), including diet-related problems. This approach proved to be fruitful for manufacturing products with optimized process parameters to achieve protein-rich cost-effective extruded snacks (Balasubramanian et al., 2012). Geeta et al. (2016) prepared extruded products from foxtail millet. Composite flour (foxtail millet, chickpea, rice, and flaxseed) is used for preparation of extruded products and their ratio of 50:15:32:3 showed the highest physical characteristics such as expansion ratio, water solubility index, water absorption index, water-holding capacity, color values, and bulk density (4.05, 0.216, 7.02 g/g, 477.6%, 79.01%(L*), and 0.074 kg/m^3, respectively). They observed that the use of composite flour (foxtail millet, rice, chickpea, and flaxseed in the ratio of 50:15:32:3) could be useful to produce quality extrudates with acceptable consumer response. Kharat et al. (2019) utilized foxtail, pearl millet, and finger millet for extrusion process. Foxtail extrudates showed the highest expansion ratio of 4.41 with a water absorption capacity of 4.18 g/g. Differences in various physical and functional properties of the individual millets affect their ability of product formulation.

8.6 PORRIDGES

In many African countries, millet is often the main component of food and is essentially consumed as steam-cooked products (couscous), thick porridges (T$_o$), and thin porridges (Ogi) that can be used as a complementary food for infants and young children. It is also used in brewing beer (Lestienne et al., 2005). Malthi et al. (2012) reported a method for finger millet porridge (Figure 8.3). Porridge consistency may be thick or thin, depending on the concentration of flour (30% down to 10%). Different types of porridges may be prepared by cooking flour in boiling water accompanied by vigorous stirring (Dias-Martins et al., 2018). Oluyimika et al. (2019)

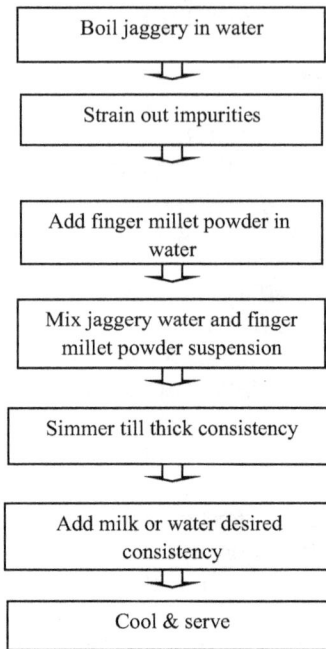

Figure 8.3 Method of preparation of finger millet porridge (Malthi et al., 2012).

evaluated the pearl millet porridge with moringa leaves and baobab fruit and added minerals. This porridge has been made to meet the shortage of minerals in food, causing mineral deficiency in people. Inclusion of baobab fruit powder, containing citric acid, improves the percentage and amount of bioaccessible iron and magnesium in pearl millet porridge, but moringa leaves have a negative effect on iron bioaccessibility. Wang et al. (2019) prepared sour porridge using broomcorn millet (proso millet). Three strains, namely, *Lactobacillus brevis* L1, *Acetobacter aceti* A1, and *Saccharomyces cerevisiae* E4, were used as inocula for the fermentation of broomcorn millet sour porridge. It is suggested that optimum fermentation conditions are the following: liquid-to-solid ratio of 3 (v/w), inoculum size of 5% (v/v) of the mixed strains as starter consisting of *L. brevis* L1, *A. aceti* A1, and *S. cerevisiae* E4 in the ratio of 1:1:1 (v/v/v), fermentation time of 30 h, and fermentation temperature of 30°C.

Mridula et al. (2015) prepared multi-grain dalia (MGD) using sprouted wheat and a mix of other three grains (barley, sorghum, and pearl millet). Protein content decreased with increase in the proportion of the three strains in MGD samples, but in vitro protein digestibility is not affected significantly. In view of very good overall sensory acceptability, rich in crude fiber, calcium, iron content and low cooking time, 25:75 parts of sprouted wheat and mixer of three grains may be considered for preparation of acceptable quality quick cooking multi-grain dalia.

8.6.1 Fura

Fura is a thick porridge consumed as a semisolid dumpling cereal meal in West Africa, particularly in Nigeria, Ghana, and Burkina Faso (Jideani et al., 2001). Fura is produced mainly from moist pearl millet flour blended with spices and compressed into balls and boiled for 30 min. Cooked dough is ground to smooth paste with the addition of water. The fura dough is rolled into a 25–30 g ball by hand and dusted with flour. The fura is made into porridge by crumbling the fura balls into fermented whole milk (kindrimo) or fermented skim milk (nono) (Jideani & Danladi, 2005). Inyang and Zakari (2008) prepared fura from pearl millet with the effect of germination and natural fermentation on the quality of fura. It was observed that the fura prepared from germinated grains had the highest value of sensory score. Anti-nutritional factor also reduced significantly after germination process. Filli and Nkama (2007) stated that millet and grain legumes (cowpea and soybean) can be used to prepare fura using HTST extrusion cooking, which revealed instantization of the product. Enhanced water absorption, hydration power, and swelling properties of extrudates are all feasible potentials to produce an instant product with improved quality.

8.6.2 Ogi

Ogi is one of the commonest food products of Nigeria, which is a white starchy mash, traditionally obtained by soaking and wet extraction of millet. "Ogi" or "akamu", a fermented cereal-based porridge, is also a popular food in Ghana, Mali, and Niger (Nkama et al., 2000). It may be mixed with boiling water to form a thin gruel called "akamu" or "ogi" porridge. Preparation of ogi includes the following steps: washing of gains, steeping for 3 days at $28 + 2°C$, wet milling, and wet sieving through pore size 300 μm, and sedimentation/souring of the filtrate for 1–3 days. Thereafter, the clean sediment (ogi) is collected after decanting water, and it is stored or consumed (Adeyemi & Beckley, 1986; Akingbala et al., 1989; Omemu & Omeike, 2010).

8.6.3 Khichri

Khichri is a traditional food in India. Rajasthani people consume more pearl millets like bajra than rice as rice is less cultivated in Rajasthan. Rice is the main ingredient of khichri in the other parts of the country. The preparation method of khichri is very easy. To prepare khichri, the grains of pearl millet are first cleaned. Then, the grains are soaked in water for 8–9 h. After, the grains are sieved to remove the excess water. Then, the soaked gains and chickpea are cooked in the pressure cooker. After that, ghee is heated in a pan and cumin seeds are added in the ghee. Then, spices such as asafetida and turmeric powder are added in the ghee and cooked on medium flame. Finally, the cooked pearl millet grains and chick-pea pulses are added, and salt is added according to taste; the mixture is then cooked on a medium flame with occasional stirring. When khichri is cooked properly, the dish is ready. The cooked khichri is served with a lot of ghee, butter, lassi, or curd. Pearl millet khichri is consumed in the winter season. Other millets can also be used for the preparation of khichri.

8.7 ALCOHOLIC AND NON-ALCOHOLIC BEVERAGES

Millet-based beverages (alcoholic and non-alcoholic) such as *malwa, bantu, pombe,* and opaque or *kaffir* beer are more prevalent as compared to other products (Adebiyi et al., 2018). Jandh (beer type) is a traditional alcoholic product (in Nepal) that is prepared using finger millet (Tamang et al., 1988). During the Jandh preparation, millet seeds are softened using steam followed by their spreading on leaves and mixing with culture (murcha). Thereafter, millet is kept in heap (24 h; ambient temperature). Next day, the mixture is placed into a pod and covered (straw and leaves). Grits are kneaded for the purpose of removing seed coat and then are mixed with water. After 10 min, the beverage is ready to consume (Karki, 1986). Kunun-zaki is one of the popular fermented non-alcoholic beverages prepared from maize, sorghum/millet, spices, and sugar (Adeyemi & Umar, 1994). Traditionally, Masvusvu is famous as sweet beverage which is prepared from malted finger millet (*Elusine coracana* (L.) Gaertn) in many villages of Zimbabwe. Mangisi is one of the sweet-sour products which is prepared using natural fermentation of sieved masvusvu (Zvauya et al., 1997). Mageu is a popular beverage which is generally consumed in South Africa and Zimbabwe (Bvochora et al., 1999; Gadaga et al., 1999). Malwa is a fermented beverage that is famous for its specific flavor, taste, and aroma. Malwa is prepared from millets at the household level in Uganda (eastern and north eastern regions) (Muyanja et al., 2010). It is served in a clay pot and consumed after diluted with hot water. Khandelwal et al. (2012) reported beverage preparation using blends of fruit juices (green and black grapes along with apple) along with germinated wheat kernels and millets (finger millet and pearl millet). They observed that the alcohol content of various blends varied from 3.3% to 6.9%. Such developments of novel low-alcohol content beverages were found to be acceptable and have good shelf life. Boza is prepared using clean water and addition of sugar followed by their fermentation. The sweet and sour tastes of boza depend on the content of acid (Arici & Daglioglu, 2002). The steps for boza production can be summarized as (1) preparation of the raw materials, (2) boiling, (3) cooling and straining, (4) sugar addition, and (5) fermentation (Amadou, 2019). *Mbege ale* is a beer made from millet, sorghum, and banana. This beer has been industrialized, packaged, and commercialized as *chibuku shake shake*. It remains a popular beer especially in Botswana, Zambia, and Zimbabwe (Adebiyi et al., 2018). *Kodo ko jaanr* is the most common fermented alcoholic beverage prepared from dry seeds of finger millet, locally called *kodo* in the eastern Himalayan regions of the Darjeeling hills and Sikkim in India, Nepal, and Bhutan (Amadou, 2019). The method for preparation of Kodo ko jaanr is shown in Figure 8.4.

8.8 LADOO

Basically, ladoo is a sweet of Indian subcontinent. It is prepared using flour, sugar, and ghee. Singh and Mehra (2017) prepared ladoo using five different combinations of Bengal gram and pearl millet flour (100, 75:25; 50:50, 25:75, 100). Incorporation of pearl millet flour above 50% was least acceptable in ladoo. At this concentration of pearl millet flour, the product became dark colored and somewhat bitter in taste as indicated by the panel, whereas inclusion of 25% millet flour gained the highest acceptability. Singh and Sehgal (2008) prepared two different ladoo using popped pearl millet. In the first type of ladoo, roasted, dehulled chickpea and groundnut were also added for the purpose to improve nutritional quality, whereas in the case of the second type of ladoo popped pearl millet (100%) was used. They observed that type I popped pearl millet ladoo had significantly higher calcium, phosphorus, and iron contents. Higher contents of polyphenol and phytic acid and lower *in vitro* protein and starch digestibility are also found in type I ladoo.

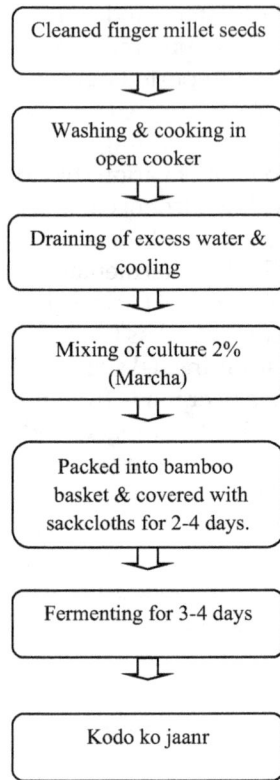

Figure 8.4   Method for preparation of Kodo ko jaanr (Amadou, 2019).

8.9 WEANING FOODS

Food composition databases constitute an essential tool for estimating the content of energy, nutrients, and other dietary compounds of the different foods. This information is necessary for the assessment of nutritional intakes and the monitoring of dietary interventions (Elmadfa & Meyer, 2010). Millets are nutritionally rich and can be utilized for preparation of various food products like weaning foods. Cereals/millets are the cheapest and most widely available source of energy; their contribution to energy intake is the highest in poor strata of society and comparatively less in the upper class. Cereals are rich sources of calcium and iron (Srivastava et al., 2015). Almeida-Dominguez et al. (1993) prepared weaning food by utilizing pearl millet and cowpea flour (70:30) with or without using malted sorghum malt. Sorghum malt hydrolyzed the starch and produced a beverage that contained 17% protein with 90% of the essential amino acids required for infants less than 1 year old. Malleshi et al. (1996) prepared weaning food by using sorghum, pearl millet, and finger millet with toasted mungbean flour and nonfat dry milk and extruded it to make ready-to-eat food. In vitro protein and carbohydrate digestibility were also evaluated, and it is observed that food prepared from sorghum, pearl millet, and finger millet had good digestibility properties. Balasubramanian et al. (2014) evaluated the weaning food prepared from extrudates of plain and malted pearl millet and barley flour. Response surface methodology was opted for development of good-quality and low-cost weaning food. They optimized the level of ingredients for weaning foods as pearl millet extrudates 20.77%, pearl millet (malt extrudates; 7.39%); barley extrudates (20.99%); barley malt extrudates (6.53%) with desirability (81.3%). Talib et al. (2017) reported weaning food preparation from pearl millet malt, wheat, and roasted rice. The formulated weaning food is reported to more dispersible, had better water-holding capacity, improved mineral content (twofold), and was more digestible as compared to wheat flour-based weaning food.

8.10 NON-DAIRY PROBIOTIC BEVERAGE

Ziemer and Gibson (1998) demonstrated that research on functional foods is moving towards dietary supplementation development along with probiotics and prebiotics concept, which ultimately affect gut microbial composition and their activities. Probiotics are living microorganisms that when administered in adequate amounts confer health benefits to the host (Hill et al., 2014). Sensory properties of nondairy probiotics are major challenges for the formulation of nondairy probiotics. The sensory properties of nondairy probiotic foods may be influenced by interactions between different probiotics strains and food substrates, where textures, taste, flavor, aroma, and color might be improved or aggravated by the production of different metabolic compounds such as lactic acid and other metabolites in living cells by different species during processing and storage (Panghal et al., 2018). Mridula and Sharma (2015) prepared a nondairy probiotic drink utilizing sprouted wheat, barley, pearl millet, and green gram separately with oat meal, stabilizer, and sugar using *L. acidophilus* NCDC14 with soymilk. It was observed that overall sensory acceptability scores for all probiotic drink samples with soymilk were higher up to 6 g of wheat, barley, and green gram, and 4 g of pearl millet flour per 100 ml liquid portion. Ziarno et al. (2019) prepared millet-based beverages using a starter containing typical lactic acid bacteria for production of yoghurt (*L. delbrueckii* subsp. *bulgaricus* and *S. thermophilus*). It was possible to prepare attractive fermented millet-based beverage. Ganguly et al. (2019) developed a probiotics beverage using whey and skim milk (60:40; v/v), flour of germinated pearl millet (4.73%; w/v), and malt extract (liquid barley; 3.27%; w/v) with *Lactobacillus acidophilus* (NCDC-13). Effect on *Shigella*-mediated mice pathogenicity was studied using these probiotics beverage. Pathogen inhibition was observed after feeding mice with probiotics beverage.

8.11 CONCLUSION

Cereals, particularly millets, have a great potential for formulation of commercial food products. Millets could be explored further to make them available for use in preparation of animal feed and baking products for humans. Millets have a great potential for food production as millets are rich sources of nutrients. Industrial applications of millet-based foods can be increased by using modern equipment and optimized conditions so that high-quality food can be prepared from millets. By using process like malting and fermentation, digestibility and nutrient content can be enhanced. Millets can be utilized in bakery, extruded, weaning, alcoholic, and non-alcoholic food products. Thus, there is a need to increase production and processing of millet.

REFERENCES

Adebiyi, J. A., Obadina, A. O., Adebo, O. A. and Kayitesi, E. 2017. Comparison of nutritional quality and sensory acceptability of biscuits obtained from native, fermented, and malted pearl millet (*Pennisetum glaucum*) flour. *Food Chemistry* 232: 210–217.

Adebiyi, J. A., Obadina, A. O., Adebo, O. A. and Kayitesi, E. 2018. Fermented and malted millet products in Africa: expedition from traditional/ethnic foods to industrial value-added products. *Critical Reviews in Food Science & Nutrition* 58: 463–474.

Adeyemi, I. A. and Beckley, T. 1986. Effect of period of maize fermentation and souring on chemical properties and amylograph pasting viscosity of ogi. *Cereal Science* 4: 353–360.

Adeyemi, T. and Umar, S. 1994. Effect of method of manufacturing of quality characteristics of kunun zaki, a millet based beverage. *Nigerian Food Journal* 12: 34–40.

Akingbala, J. O., Rooney, L. W. and Faubion, J. M. 1989. A laboratory procedure for the preparation of ogi, a Nigerian fermented food. *Journal of Food Science* 46 (5): 1523–1526.

Almeida-Dominguez, H. D., Serna-Saldivar, S. O., Gomez, M. H. and Rooney, L. W. 1993. Production and nutritional value of weaning foods from mixtures of pearl millet and cowpeas. *Cereal Chemistry* 70: 14–18.

Amadou, I. 2019. Millet based fermented beverages processing. In: Grumezescu, A. M. and Holban, A. M., editor. *Fermented beverages*. UK: Woodhead Publishing. pp. 433–472.

Annor, G. A., Tyl, C., Marcone, M., Ragaee, S. and Marti, A. 2017. Why do millets have slower starch and protein digestibility than other cereals? *Trends in Food Science & Technology* 66: 73–83.

Arici, M. and Daglioglu, O. 2002. Boza: a lactic acid fermented cereal beverage as a traditional Turkish food. *Food Reviews International* 18: 39–48.

Awolu, O. O. 2017. Optimization of the functional characteristics, pasting and rheological properties of pearl millet-based composite flour. *Heliyon* 3: e00240.

Awolu, O. O., Olarewaju, O. A. and Akinade, A. O. 2017. Effect of the addition of pearl millet flour subjected to different processing on the antioxidants, nutritional, pasting characteristics and cookies quality of rice based composite flour. *Journal of Nutritional Health & Food Engineering* 7: 00232.

Balasubramanian, S., Kaur, J. and Singh, D. 2014. Optimization of weaning mix based on malted and extruded pearl millet and barley. *Journal of Food Science and Technology* 51: 682–690.

Balasubramanian, S., Singh, K. K., Patil, R. T. and Onkar, K. K. 2012. Quality evaluation of millet-soy blended extrudates formulated through linear programming. *Journal of Food Science and Technology* 49: 450–458.

Belorio, M. and Gómez, M. 2020. Gluten-free muffins versus gluten containing muffins: ingredients and nutritional differences. *Trends in Food Science and Technology* 102: 249–253.

Bhol, S. and Bosco, S. J. D. 2014. Influence of malted finger millet and red kidney bean flour on quality characteristics of developed bread. *LWT-Food Science and Technology* 55(1): 294–300.

Bvochora, J. M., Reed, J. D., Read, J. S. and Zvauya, R. 1999. Effect of fermentation processes on proanthocyanidins in sorghum during preparation of Mahewu, a non-alcoholic beverage. *Process Biochemistry* 35: 21–25.

Chandrasekara, A. and Shahidi, F. 2010. Content of insoluble bound phenolics in millets and their contribution to antioxidant capacity. *Journal of Agricultural and Food Chemistry* 58: 6706–6714.

213

Chauhan, E. S. and Sarita 2018. Effects of processing (germination and popping) on the nutritional and anti-nutritional properties of finger millet (*Eleusine Coracana*). *Current Research in Nutrition & Food Science Journal* 6: 566–572.

Choudhury, M., Das, P. and Baroova, B. 2011. Nutritional evaluation of popped and malted indigenous millet of Assam. *Journal of Food Science & Technology* 48: 706–711.

Cordelino, I. G., Tyl, C., Inamdar, L., Vickers, Z., Marti, A. and Ismail, B. P. 2019. Cooking quality, digestibility, and sensory properties of proso millet pasta as impacted by amylose content and prolamin profile. *LWT-Food Science & Technology* 99: 1–7.

Dias-Martins, A. M., Pessanha, K. L. F., Pacheco, S., Rodrigues, J. A. S. and Carvalho, C. W. P. 2018. Potential use of pearl millet (*Pennisetum glaucum* (L.) R. Br.) in Brazil: food security, processing, health benefits and nutritional products. *Food Research International* 109: 175–186.

Ebba, T. 1969. T'ef (Eragrostis tef). The cultivation, usage and some of the known diseases and insect pests. Experimental Station Bulletin. No. 60. Haile Sellassie I University, College of Agriculture: Dire Dawa, Ethiopia.

Elmadfa, I. and Meyer, A. L. 2010. Importance of food composition data to nutrition and public health. *European Journal of Clinical Nutrition* 64: S4–S7.

Filli, K. B. and Nkama, I. 2007. Hydration properties of extruded fura from millet and legumes. *British Food Journal* 109: 68–80.

Gadaga, T. H., Mutukumira, A. N., Narvhus, J. A. and Feresu, S. B. 1999. A review of traditional fermented foods and beverages of Zimbabwe. *International Journal of Food Microbiology* 53: 1–11.

Ganguly, S., Sabikhi, L. and Singh, A. K. 2019. Effect of whey-pearl millet-barley based probiotic beverage on Shigella-induced pathogenicity in murine model. *Journal of Functional Foods* 54: 498–505.

Geeta, H. P., Mathad, P., Nidoni, U. and Ramachandra, C. T. 2016. Development of foxtail millet based extruded food product. *International Journal of Food Science & Technology* 6: 11–22.

Goesaert, H., Leman, P., Bijttebier, A. and Delcour, J. A. 2009. Antifirming effects of starch degrading enzymes in bread crumb. *Journal of Agricultural & Food Chemistry* 57: 2346–2355.

Goswami, D., Gupta, R. K., Mridula, D., Sharma, M. and Tyagi, S. K. 2015. Barnyard millet based muffins: physical, textural and sensory properties. *LWT-Food Science & Technology* 64: 374–380.

Gull, A., Prasad, K. and Kumar, P. 2015. Effect of millet flours and carrot pomace on cooking qualities, color and texture of developed pasta. *LWT-Food Science & Technology* 63: 470–474.

Hill, C., Guarner, F., Reid, G., Gibson, G. R., Merenstein, D. J., et al. 2014. The International Scientific Association for Probiotics and Prebiotics consensus statement on the scope and appropriate use of the term probiotic. *Nature Reviews Gastroenterology & Hepatology* 11(8): 506–514.

Inyang, C. U. and Zakari, U. M. 2008. Effect of germination and fermentation of pearl millet on proximate, chemical and sensory properties of instant "Fura"-a Nigerian cereal food. *Pakistan Journal of Nutrition* 7: 9–12.

Jalgaonkar, K. and Jha, S. K. 2016. Influence of particle size and blend composition on quality of wheat semolina-pearl millet pasta. *Journal of Cereal Science* 71: 239–245.

Jalgaonkar, K., Jha, S. K. and Mahawar, M. K. 2018. Influence of incorporating defatted soy flour, carrot powder, mango peel powder, and moringa leaves powder on quality characteristics of wheat semolina-pearl millet pasta. *Journal of Food Processing & Preservation* 42: 13575.

Jaybhaye, R. V., Pardeshi, I. L., Vengaiah, P. C. and Srivastav, P. P. 2014. Processing and technology for millet based food products: a review. *Journal of Ready to Eat Food* 1: 32–48.

Jideani, V. A. and Danladi, I. M. 2005. Instrumental and sensory textural properties of fura made from different cereal grains. *International Journal of Food Properties* 8: 49–59.

Jideani, V. A., Nkama, I., Agbo, E. B. and Jideani, I. A. 2001. Survey of fura production in some northern states of Nigeria. *Plant Foods for Human Nutrition* 56: 23–36.

Karki, 1986. Some Nepalese fermented food and beverages: traditional food, some products and technologies. Central Food Technology and Research Institute, Mysure, India.

Kaur, P., Purewal, S. S., Sandhu, K. S., Kaur, M. and Salar, R. K. 2019. Millets: a cereal grain with potent antioxidants and health benefits. *Journal of Food Measurement & Characterization* 13: 793–806.

Khandelwal, P., Upendra, R. S., Kavana, U. and Sahithya, S. 2012. Preparation of blended low alcoholic beverages from under-utilized millets with zero waste processing methods. *International Journal of Fermented Foods* 1: 77–86.

Kharat, S., Medina-Meza, I. G., Kowalski, R. J., Hosamani, A., Ramachandra, C. T., Hiregoudar, S. and Ganjyal, G. M. 2019. Extrusion processing characteristics of whole grain flours of select major millets (foxtail, finger, & pearl). *Food & Bioproducts Processing* 114: 60–71.

Krishnan, R., Dharmaraj, U., Manohar, R. S. and Malleshi, N. G. 2011. Quality characteristics of biscuits prepared from finger millet seed coat based composite flour. *Food chemistry* 129: 499–506.

Kulkarni, S. K., Sakhale, B. K., Pawar, V. D., Miniyar, U. G. and Patil, B. M. 2010. Studies on sensory quality of cookies enriched with mushroom powder. *Food Science Research Journal* 1: 90–93.

Kulthe, A. A., Thorat, S. S. and Khapre, A. P. 2018. Nutritional and sensory characteristics of cookies prepared from pearl millet flour. *The Pharma Innovation Journal* 7(4): 908–913.

Lestienne, I., Mouquet-Rivier, C., Icard-Verniere, C., Rochette, I. and Treche, S. 2005. The effects of soaking of whole, dehulled and ground millet and soybean seeds on phytate degradation and Phy/Fe and Phy/Zn molar ratios. *International Journal of Food Science & Technology* 40: 391–399.

Malleshi, N. G. and Desikachar, H. S. R. 1981. Varietal differences in puffing quality of ragi (*Elusine coracana*). *Journal of Food Science & Technology* 26: 26–28.

Malleshi, N. G., Hadimani, N. A., Chinnaswamy, R. and Klopfenstein, C. F. 1996. Physical and nutritional qualities of extruded weaning foods containing sorghum, pearl millet, or finger millet blended with mung beans and nonfat dried milk. *Plant Foods for Human Nutrition* 49: 181–189.

Malthi, D., Sindhumathi, G. and Thilagavathi, T. 2012. Traditional recipes from finger millet. https://www.dhan.org/smallmillets/docs/books/Receipe_ booklet_finger_millet.pdf (Downloaded 22.05.2020).

McSweeney, M. B., Ferenc, A., Smolkova, K., Lazier, A., Tucker, A., Seetharaman, K., Wright, A., Duizer, L. M. and Ramdath, D. D. 2017. Glycaemic response of proso millet-based (*Panicum miliaceum*) products. *International Journal of Food Sciences & Nutrition* 68(7): 873–880.

Mildner-Szkudlarz, S., Bajerska, J., Gornas, P., Seglina, D., Pilarska, A. and Jesionowski, T. 2016. Physical and bioactive properties of muffins enriched with raspberry and cranberry pomace powder: a promising application of fruit by-products rich in biocompounds. *Plant Foods for Human Nutrition* 71: 165–173.

Misra, N. N. and Tiwari, B. K. 2014. Biscuits. In: Zhou, W., Hui, Y. H., De Leyn, I., Pagani, M. A., Rosell, C. M., Selman, J. D. and Therdthai, N., editors. *Bakery products science and technology*. West Sussex, UK: John Wiley & Sons, Ltd. pp. 585–601.

Moore, G. 1994. Snack food extrusion. In Frame, N. D., editor. *The technology of extrusion cooking*. Bishopbriggs, Glasgow, UK: Blackie Academic and Professional, an imprint of Chapman and Hall. pp. 110–143.

Mridula, D. and Sharma, M. 2015. Development of non-dairy probiotic drink utilizing sprouted cereals, legume and soymilk. *LWT - Food Science Technology* 62: 482–487.

Mridula, D., Sharma, M. and Gupta, R. K. 2015. Development of quick cooking multi-grain dalia utilizing sprouted grains. *Journal of Food Science & Technology* 52: 5826–5833.

Muyanja, C., Birungi, S., Ahimbisibwe, M., Semanda, J. and Namugumya, B. S. 2010. Traditional processing, microbial and physicochemical changes during fermentation of malwa. *African Journal of Food Agriculture Nutrition & Development* 10: 4124–4138.

Nkama, I., Dappiya, S., Modu, S. and Ndahi, W. 2000. Physical, chemical, rheological and sensory properties of Akama from different pearl millet cultures. *Journal Aridland Agriculture* 10: 145–149.

Oluyimika, Y. A., Kruger, J., White, Z. and Taylor, J. R. 2019. Comparison between food-to-food fortification of pearl millet porridge with moringa leaves and baobab fruit and with adding ascorbic and citric acid on iron, zinc and other mineral bioaccessibility. *LWT-Food Science & Technology* 106: 92–97.

216

Omemu, A. M. and Omeike, S. O. 2010. Microbiological hazard and critical control points identification during household preparation of cooked ogi used as weaning food. *International Food Research Journal* 17: 257–266.

Onyango, C., Luvitaa, S. K., Unbehend, G. and Haase, N. 2020. Physico-chemical properties of flour, dough and bread from wheat and hydrothermally-treated finger millet. *Journal of Cereal Science* 93: 102954.

Panghal, A., Janghu, S., Virkar, K., Gat, Y., Kumar, V. and Chhikara, N. 2018. Potential non-dairy probiotic products–A healthy approach. *Food Bioscience* 21: 80–89.

Panghal, A., Khatkar, B. S. and Singh, U. 2006. Cereal proteins and their role in food industry. *Indian Food Industry* 25: 58–62.

Panghal, A., Khatkar, B. S., Yadav, D. N. and Chhikara, N. 2019. Effect of finger millet on nutritional, rheological, and pasting profile of whole wheat flat bread (chapatti). *Cereal Chemistry* 96: 86–94.

Pareyt, B. and Delcour, J. A. 2008. The role of wheat flour constituents, sugar, and fat in low moisture cereal based products: a review on sugar-snap cookies. *Critical Reviews in Food Science and Nutrition* 48: 824–839.

Parker, M. L., Umeta, M. and Faulks, R. M. 1989. The contribution of flour components to the structure of injera, an Ethiopian fermented bread made from tef (Eragrostis tef). *Journal of Cereal Science* 10(2): 93–104.

Purewal, S., Salar, R., Bhatti, M., Sandhu, K. S., Singh, S. K. and Kaur, P. 2020. Solid-state fermentation of pearl millet with *Aspergillus oryzae* and *Rhizopus azygosporus*: effects on bioactive profile and DNA damage protection activity. *Journal of Food Measurement & Characterization* 14: 150–162.

Purewal, S. S., Sandhu, K. S., Salar, R. K. and Kaur, P. 2019. Fermented pearl millet: a product with enhanced bioactive compounds and DNA damage protection activity. *Journal of Food Measurement & Characterization* 13: 1479–1488.

Rajiv, J., Soumya, C., Indrani, D. and Venkateswara Rao, G. 2011. Effect of replacement of wheat flour with finger millet flour (*Eleusine corcana*) on the batter microscopy, rheology and quality characteristics of muffins. *Journal of Texture Studies* 42: 478–489.

Ravindran, G. 1992. Seed protein of millets: amino acid composition, proteinase inhibitors and *in-vitro* protein digestibility. *Food Chemistry* 44: 13–17.

Reilly, N. R. 2016. The gluten-free diet: recognizing fact, fiction, and fad. *The Journal of Pediatrics*: 175.

Saha, S., Gupta, A., Singh, S. R. K., Bharti, N., Singh, K. P., Mahajan, V. and Gupta, H. S. 2011. Compositional and varietal influence of finger millet flour on rheological properties of dough and quality of biscuit. *LWT-Food Science & Technology* 44: 616–621.

Salar, R. K. and Purewal, S. S. 2017. Phenolic content, antioxidant potential and DNA damage protection of pearl millet (*Pennisetum glaucum*) cultivars of North Indian region. *Journal of Food Measurement & Characterization* 11: 126–133.

Saleh, A. S., Zhang, Q., Chen, J. and Shen, Q. 2013. Millet grains: nutritional quality processing, and potential health benefits. *Comprehensive Reviews in Food Science & Food Safety* 12: 281–295.

Sarabhai, S., Tamilselvan, T. and Prabhasankar, P. 2020. Role of enzymes for improvement in gluten-free foxtail millet bread: IT'S effect on quality, textural, rheological and pasting properties. *LWT-Food Science & Technology*: 110365.

Sharma, B. and Gujral, H. S. 2019. Modulation in quality attributes of dough and starch digestibility of unleavened flat bread on replacing wheat flour with different minor millet flours. *International Journal of Biological Macromolecules* 141: 117–124.

Sharma, S., Saxena, D. C. and Riar, C. S. 2016. Nutritional, sensory and in-vitro antioxidant characteristics of gluten free cookies prepared from flour blends of minor millets. *Journal of Cereal Science* 72: 153–161.

Singh, A. and Kumar, P. 2018. Optimization of gluten free biscuit from foxtail, copra meal and amaranth. *Food Science & Technology* 39: 43–49.

Singh, U. and Mehra, A. 2017. Sensory evaluation of Ladoo prepared with pearl millet. *International Journal of Home Science* 3: 610–612.

Singh, G. and Sehgal, S. 2008. Nutritional evaluation of ladoo prepared from popped pearl millet. *Nutrition & Food Science* 38: 310–315.

Siroha, A. K., Sandhu, K. S. and Kaur, M. 2016. Physicochemical, functional and antioxidant properties of flour from pearl millet varieties grown in India. *Journal of Food Measurement & Characterization* 10: 311–318.

Srivastava, S. and Batra, A. 1998. Popping qualities of minor millets and their relationship with grain physical properties. *Journal of Food Science & Technology* 35: 265–267.

Srivastava, S., Neerubala, S. S. and Shamim, M. Z. 2015. Nutritional composition of weaning food using malted cereal and pulses flour for infants. *International Journal of Pure & Applied Bioscience* 3: 171–185.

Sullivan, J. F. and Craig Jr, J. C. 1984. The development of explosion puffing. *Food technology (USA)*.

Talib, M. I., Burse, A. and Parate, V. R. 2017. Development of weaning food from pearl millet malt. *International Conference Proceeding*, 429–434. http://doi.one/10.1727/IJCRT.17174.

Tamang, J. P., Sarkar, P. K. and Hesseltine, C. W. 1988. Traditional fermented foods and beverages of Darjeeling and Sikkim–a review. *Journal of the Science of Food & Agriculture* 44: 375–385.

Tebben, L., Shen, Y. and Li, Y. 2018. Improvers and functional ingredients in whole wheat bread: a review of their effects on dough properties and bread quality. *Trends in Food Science & Technology* 81: 10–24.

Taylor, J. R. N. and Emmanbux, M. N. 2008. Millets. In: Arend, A. and Bello, F. D., editors. *Handbook of gluten free cereal*. US: Academic Press.

Tomić, J., Torbica, A. and Belović, M. 2020. Effect of non-gluten proteins and transglutaminase on dough rheological properties and quality of bread based on millet (*Panicum miliaceum*) flour. *LWT-Food Science & Technology* 118: 108852.

Wang, Q., Liu, C., Jing, Y. P., Fan, S. H. and Cai, J. 2019. Evaluation of fermentation conditions to improve the sensory quality of broomcorn millet sour porridge. *LWT-Food Science & Technology* 104: 165–172.

Xu, J., Wang, W. and Li, Y. 2019. Dough properties, bread quality, and associated interactions with added phenolic compounds: a review. *Journal of Functional Foods* 52: 629–639.

Xu, J., Zhang, Y., Wang, W. and Li, Y. 2020. Advanced properties of gluten-free cookies, cakes, and crackers: a review. *Trends in Food Science & Technology* 103: 200–213.

Ziarno, M., Zaręba, D., Henn, E., Margas, E. and Nowak, M. 2019. Properties of non-dairy gluten-free millet-based fermented beverages developed with yoghurt cultures. *Journal of Food & Nutrition Research* 58: 21–30.

Ziemer, C. J. and Gibson, G. R. 1998. An overview of probiotics, prebiotics and synbiotics in the functional food concept: perspectives and future strategies. *International Dairy Journal* 8: 473–479.

Zvauya, R., Mygochi, T. and Parawira, W. 1997. Microbial and biochemical changes occurring during production of masvusvu and mangisi, traditional Zimbabwean beverages. *Plant Foods for Human Nutrition* 51: 43–51.

9

Millet Diseases and Their Control

9.1 INTRODUCTION

Diseases in millets have significant effects as they result in economic as well as postharvest losses in the grain production sector throughout the world. Detection of disease-causing pathogens in millets is crucial for the maintenance of sustainability in the farming sector. Using eco-friendly methods/technologies that are non-destructive in nature is one of the important practical and feasible ways for evaluating the health status of millets for their industrial applications. Detection of diseases at an earlier stage could help the farmer save their crops from the damaging effects of pathogens. The methods currently being used for disease detection include morphological analysis and serological and molecular techniques, which are either costly, time-consuming, or need complex practical protocols. Hence, the necessity of nondestructive, cost-effective evaluation methods arises, as they could ease the disease detection methods. Prevention of microorganism-based diseases in millets during early stages helps in maintaining millet yield per hectare; at the same time, the process decreases pesticide dependence. To overcome the disease conditions in millets, it is necessary to understand the disease-causing pathogen, their life cycle, and factors that play crucial roles in the pathogenesis. The analysis helps to control the disease spread as early as possible.

221

Proper management during pre- and postharvest periods helps the farmers in getting optimal yield from their field as well as keeping the parameters of millet grains fit for consumer demand.

Improvement and enhancement of crop production using specifically designed techniques and protocols is of utmost importance. Improved agricultural output fulfills the need of hunger and also provides pharmaceutically important secondary metabolites, wood, gums, and fibers (Kaur & Purewal, 2019; Kaur et al., 2018; Herve et al., 2016). Plant diseases are one of the major causes of decline in agricultural outputs. They are characterized by certain specific symptoms within the plant system and on the surface of plants (Satya & Sarkar, 2018; Baltes et al., 2017; Savary et al., 2012). Occurrence of diseases in plants results in irregularity in specific functionality of cellular reaction mechanisms, which ultimately affect the growth and development process (Andersen et al., 2018; Hückelhoven, 2007; Niks & Rubiales, 2002). This abnormal functionality serves as a major constrain in the production system. The severity of disease decides the loss to the production system as, sometimes, disease induces changes in the production, which is accepted due to lack of effective control or disease detection protocols (Cunniffe et al., 2014; Juroszek & Von-Tiedemann, 2011; Shcherbakova, 2007). Major factors that play vital roles in the occurrence of diseases in crops are temperature, rainfall, soil profile, pH, moisture, growth stage of plants, wind, deficiency of nutrients, weak immune system, and chemical-induced damages (Abdou Zayan, 2020; Velásquez et al., 2018; Jahromi, 2007). Under favorable conditions, opportunistic microorganisms abound and cause diseases in crops. More than 85% of the diseases in crops are caused by fungal strains, followed by bacteria and viruses.

Crop diseases generate the necessity of effective agricultural practices, use of fertilizers and fungicides, and use of disease-resistant varieties of crops. Genetically improved crop varieties are less vulnerable to diseases caused by biotic and abiotic factors (Kumar et al., 2020; Van Esse et al., 2019; Verma, 2013). With renewed scientific technologies and great emphasis on intensive management, the disease-related problems in cropping systems could be solved. Disease occurrences in crops can be effectively controlled by understanding and studying the major reasons that favor the disease occurrences followed by their control. The control measures also depend on the timing of diseases in crops (pre- or postharvest). To minimize the loss to crops due to fungal strains, rust, and bad weather, it is necessary to harvest them on time. To avoid diseases in fresh grains,

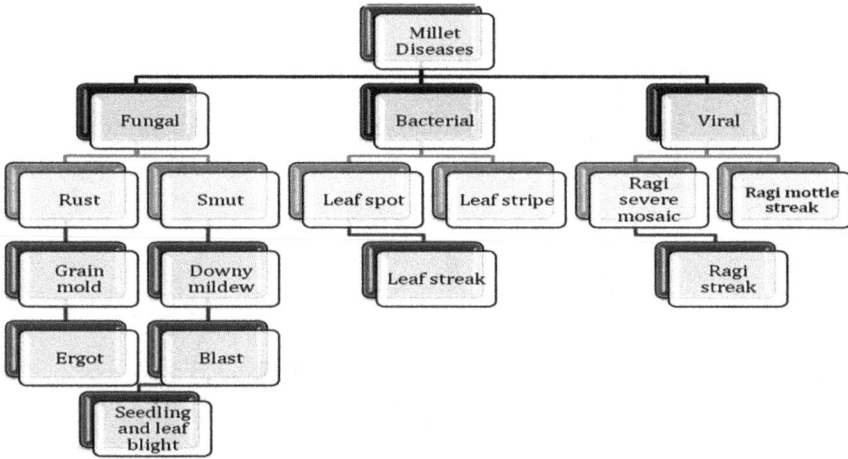

Figure 9.1 Millet diseases caused by different microorganisms.

they should go through sequential steps. Postharvest management skills determine the overall success for long-term storage of crops. This chapter describes millet diseases caused by microbial strains (fungi, bacteria, and virus), their control, and pre- and postharvest management. Various millets diseases are reported in Figure 9.1 and the microorganisms, preferred hosts, and disease control strategies are briefly reported in Table 9.1. This chapter provides an in-depth information on millet diseases, symptoms, and their effective control measures.

9.2 FUNGAL DISEASES IN MILLETS

Fungal strains are causing significant losses in millet crops as they have the potential to survive even during harsher conditions of environment. The presence of even minimal moisture content in the field helps them to survive and maintain their life cycle. Fungal strains produce diseases in millets either through their direct growth on millet plants or through secondary hosts. Air, humidity, temperature, and growth stage of millets play crucial roles in disease spread. Various fungal strain-mediated diseases are explained in the following.

223

Table 9.1 Brief Description of Diseases

Diseases	Causative microorganism	Preferred host	References
Rust	*Puccinia substriata*	Pearl millet and small millets	Nyvall (1989); Rai and Thakur (1995); Dang and Panwar (2004); De-Carvalho et al., 2006; Saveetha et al. (2007); Sharma et al. (2013); Prakash et al. (2014); Prakash et al. (2019)
Smut	*Melanopsichium eleusinis, Ustilago crameri, Ustilago panici-frumentacei, Sorosporium paspali thunbergii*	Finger millet, foxtail millet and barnyard millet, kodo and proso millets	
	Aspergillus spp., C. lunata, Fusarium spp., Bipolaris spp., Alternaria alternate and Phoma sorghina	Finger millet and pearl millet	
Downy mildew	*Sclerospora graminicola*	Pearl millet and small millets	
Ergot	*C. fusiformis*	Pearl millet	
Blast	*Pyricularia grisea*	Pearl millet, finger millet, foxtail millet, proso millet, small millets and barnyard millet	
Seedling and leaf blast	*Drechslera nodulosum Berk and Curt.*	Millets	
Leaf spot	*P. syringae, Xanthomonas eleusinae*	Finger millet and pearl millet	
Leaf stripe	*Pseudomonas avenae, Pseudomonas eleusinae*	Finger millet and pearl millet	
Leaf streak	*Xanthomonas axonopodis pv. Pennamericanum, Xanthomonas axonopodis pv. Coracanae, Pseudomonas avenae*	Pearl millet, small millets, foxtail, barnyard, proso millets	
Ragi severe mosaic	*Sugarcane mosaic virus*	Minor millets	
Ragi mottle streak	*Rhabdovirus*	Minor millets	
Ragi streak	*Eleusine strain of maize streak virus*	Minor millets	

9.2.1 Rust

Disease causing microorganisms: *Puccinia substriata*
  Specific host: Pearl millet and small millets

Major symptoms

Rust is a kind of fungal infection in millets that results in reduction in grain yield and deterioration of quality parameters. The percentage of loss may vary with the symptoms and stages of crop growth during the onset of fungal infections. The disease may spread to a larger portion of the field through air passage and the spores of causal microorganism that have the capability to survive even in soil, plant debris, and alternate hosts. The causal organism shows a complex life cycle, divided into different phases with the requirement of two alternative hosts with no relationship among them (Figure 9.2).

Teliospores start their germination and produce aeciospores, the main cause of infection in eggplants. The sexual reproduction phase is performed in eggplants. Mating results in the formation of yellow-colored pustules, which, with time, turns reddish brown. Cup-shaped protruding structures with aeciospores finally infect millet crops, and here the life cycle completes. Initially, symptoms start with the appearance of yellowish-white spots on both upper and lower side of leaf. With time,

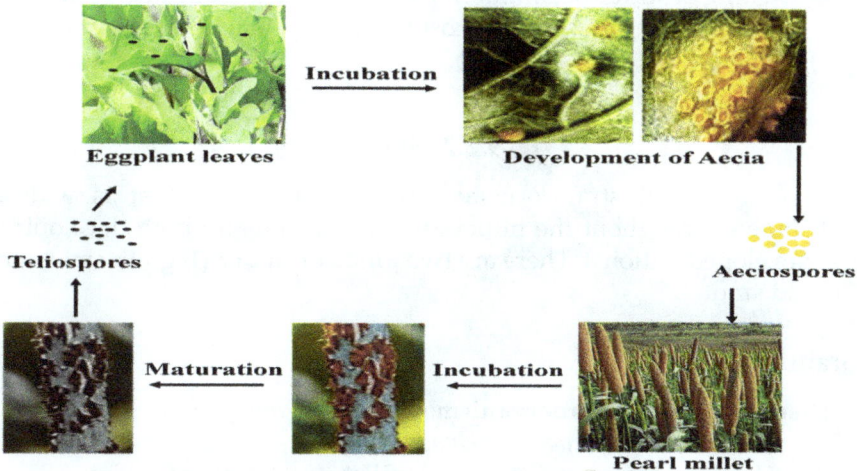

Figure 9.2 Rust progression cycle.

the spots merge and result in the formation of red/orange-colored rusty pustules with yellow margins that ultimately turn darker with the progression of disease. Depending on the infection severity and plant growth stage, the decaying of leaf starts and, in rare case, the whole plant may collapse. A unique feature of rust disease is that it can infect at any stage; however, if the plant becomes infected before the flowering stage, the risk of crop loss becomes high. The conditions that favor the progression of disease are warm temperatures during the day of 25°C–34°C and low temperatures at night of 15°C–20°C. These conditions favor the growth of dew-like droplets on leaves.

Control

- Use rust-resistant millet cultivars.
- Early sowing may reduce to loss to crop.
- Avoid the growth of disease-favoring crops (tomato, potato, and pepper).
- Control the growth of unwanted plants nearby to main crop.
- Remove the plant remnant/debris so that chances of fungal infection could be minimized.
- Crop rotation could prove to be a beneficial step.
- Use suitable fungicides.
- Dusting of sulfur or similar inorganic chemicals may help to minimize the risk; however, the cost is somewhat higher for marginal farmers.

9.2.2 Smut

Smut is a fungal strain-oriented disease of millets that is widely distributed throughout the millet-growing regions (in both developing and developed nations). There are two kinds of smuts: (1) grain smut and (2) head smut.

Grain smut

Disease-causing microorganisms: *Melanopsichium eleusinis; Ustilago crameri; Ustilago panici-frumentacei*
Specific host: Finger millet, foxtail millet, and barnyard millet

Head smut

Disease-causing microorganisms: *Sorosporium paspalithunbergii*
Specific host: Barnyard, Kodo, and Proso millets

Major symptoms
During the smut disease in millets, ovaries in the infected part (florets) start converting into sori, which is characterized by their size, usually 3–4 mm long and broader at top (2–3 mm). During the initial stage of infection, bright green-colored sori evolves, which converts slowly to dark brown to black. In this disease, grains are replaced by dense mass of black spores covered with thin film. As the disease moves to the next phase (maturity), the brownish-black-colored spores start bursting to the surrounding regions, ultimately causing infection in healthy plants. Under favorable conditions (pH, temperature, and moisture), sporeballs germinate to form networks of mycelia, which further proceeds to teliospore followed by sporidia formation, ultimately infecting the host plant before flowering. The causative agent (fungal strain) is capable of growing under a wider temperature range (Purdy & Kendrick, 1963; Mahdi, 1962).

Control

- Use of effective fungicides (Ceresan, Agrosan, Zineb, Mancozeb, Plantvax, Vitavax, and Benlate) could be promising to control the symptoms of smut in millets (Thakur & King, 1988).
- Growing smut-resistant cultivars (WC-C75, ICMV 155, ICMS 7703, and ICTP 8203) in the field helps to eradicate the smut problem.
- Before sowing, pre-treatment of seeds with Carbendazim/Thiram (@2 g/kg) helps avoid this problem.

9.2.3 Grain Mold

Disease-causing microorganisms: *Aspergillus* spp.; *C. lunata*; *Fusarium* spp.; *Bipolaris* spp.; *Alternaria alternate*; and *Phoma sorghina*
  Specific host: Finger millet and pearl millet

Major symptoms

- Pigmentation of lemma
- Growth of fungus on filaments and anther

- Poor seed set
- Shriveled grains

Mold infections in millet grains start with the growth of colored filaments on grains. Further enhanced fungal growth may reach the internal tissues of grains and result in damage at the internal level. Diseases result in decreased production and deterioration of quality of agricultural products (Frederiksen et al., 1982; ICRISAT, 1987). The severity of mold-assisted infections in millets may vary with the season as well as storage conditions (temperature and moisture) and growth stage of grains. Navi et al. (2005) reported that molds have the potential to grow within the grains and the infection may spread via contact, poor conditions during transport, and storage. The symptoms of diseases may vary from cultivar to cultivar, and environment conditions act as a determinantal factor. Mold infection in grains deteriorates/degrades the grain quality parameters, causes loss of seed mass, decreases the density of grains, and, ultimately, affects the processing of grains at industrial level. Through the food chain, the mold infection may infect animals and even humans through fungal spore inhalation/hypersensitive allergic reactions (Somani & Indira, 1999; Indira & Rana, 1997). Some fungal strains are involved in the production of mycotoxins such as trichothecenes, aflatoxins, zearalenone, and fumonisins (Bilgrami & Choudhary, 1998; Bhat et al., 1997).

Control: Using mold-tolerant cultivars, harvesting crop at appropriate stage, and maintaining proper storage conditions

9.2.4 Downy Mildew

Disease-causing microorganism: *Sclerospora graminicola*
Specific host: Pearl millet and small millets

Major symptoms

- Systemic/localized infection
- Infected seedlings become pale yellow in color
- Disease initially starts from leaf blade (lower part)
- Lower part of leaf is covered with white conidia/conidiophores
- Infected area shows necrosis/chlorosis, turns brownish in color, and ultimately disintegrates
- Rectangular-shaped lesions on infected site

Pearl millet is an important crop belonging to the *Poaceae* family and being grown by farmers throughout the world due to its industrial values (Salar & Purewal, 2017, 2016; Siroha et al., 2016; Jurjevic et al., 2007). Downy mildew is one of the common diseases occurring in pearl millet and small millets. In the beginning, white colored conidia starts their growth on lower parts of younger leaves, which is an indication of oospores formation. Oospores act as disease reservoirs and remain actively viable for a period of 10 years (Prakash et al., 2014; Nene & Singh, 1976). Infection proceeds through radicals, lower parts of coleoptiles, and stem bases (underground portions). The presence of high humidity, rainfall, and low temperature works as determinantal factors for the progression of disease symptoms in millets. Plants appear stunted and may die prematurely due to the diseased conditions. Plants infected with downy mildew show weak immunity and become susceptible to the onset of other infectious diseases (Williams, 1984).

Control
The disease-causing microorganism (obligate biotroph) may be soil-/air- or seed-borne with the capability to grow in restricted range, i.e., living tissues. Infection in crops starts during their vegetative phase; hence, the controlling measures should be initiated from the starting seedling stage. An effective way to control the disease condition is the use of disease-resistant cultivars. Further, an effective fungicidal spray may prove to be beneficial.

9.2.5 Ergot

Disease-causing microorganism: *C. fusiformis*
Specific host: Pearl millet

Major symptoms

- Secretion of viscous honeydew-like droplets from infected parts
- Formation of sclerotium (wart-like fungal structure)
- Growth of fungi in infected area usually with black coloration
- Instead of grains, fungi start to grow within the infected area

Ergot is a fungus-assisted infection that mainly affects the grain-bearing heads in pearl millet. One of the first symptoms during ergot disease is the secretion of honeydew-like creamish-pink-colored viscous droplets from the flowering buds. Onset of ergot in pearl millet resulted in reduction of grain yield as well as deterioration of grain quality. Further, the infection may pose health hazards to organisms that rely on pearl millet for food and feed. Transmission route of ergot disease may be soil-/air- or seed-borne spores. Air and heavy rainfall mediate the transfer of spores to pearl millet, and disease severity may enhance with the duration of time and growth stage.

Control

- Ergot disease in pearl millet could be managed by using resistant varieties of pearl millet.
- Selection of pearl millet cultivars that specifically enters in flowering stage during dry weather.
- Removal of infected parts from the plants so as to restrict the infection only to the limited area.
- Use of broad-spectrum fungicides.

9.2.6 Blast

Disease-causing microorganism: *Pyricularia grisea*

Specific host: Pearl millet, Finger millet, foxtail millet, proso millet, small millets, and barnyard millet

The major hosts for blast infections are cereals including wheat, rice, and millets, causing damage at a major level (Pennisi, 2010). One of the most affected crops by the fungal strain is millets (Nakayama et al., 2005). Nagaraja and Mantur (2007) demonstrated that in the severe form, fungal infection may cause 30%–40% reduction in grain yield. The causative organisms initially pose infection in crops via air passage or transfer from infected weeds. Blast-causing fungi are capable enough to target peduncles, fingers, leaves, and even seedlings. Depending on the growth stage of crop, the severity of fungal infections varies accordingly. Initial symptoms of blast disease are the appearance of diamond-/elliptical-shaped lesions with a grey center on the leaf. Under favorable conditions, spots may grow further and infected area starts getting a blasted look. During the disease, the most critical stage is neck infection, which results in loss of grains and poor-quality grain development with increased sterility in spikelet. During high-humid conditions, lesions start producing spores that infect other healthy plants. Extensive loss of chlorophyll from infected leaves resulted in the death of young leaves.

Control

- Use of blast-resistant millet cultivars (GPU-26, GPU-28, GPU-48)
- Pretreatment of seeds with suitable fungicides
- Two sprays of Saaf (0.2%)/carbendazim (0.05%)/tricyclazole 0.05%

9.2.7 Seedling Blight and Leaf Blight

The causative agent for the onset of seedling and leaf blight in millets is *Drechslera nodulosum* Berk and Curt. The disease causes loss to crop yield with a significant economic damage. Butler (1918) observed the onset of disease from different regions of India. At present, the disease is widely distributed in the Philippines, Africa, Japan, US, Uganda, and east African regions. During high humidity in the environment, fungal strains start to grow on older parts of plants, which ultimately showed symptoms, i.e., appearance of brown-colored spots. Depending on the severity of disease, the fungal infection may result in discoloration of seeds. Fungal strains may remain viable even on plant debris, soil, and stubble. The optimal temperature conditions for the spread of fungal infections are 30°C–32°C. The infection may spread to healthy plants through air passage, and fungal spores have viability up to 1 year.

9.3 BACTERIAL DISEASES IN MILLETS

9.3.1 Leaf Spot

Disease-causing microorganisms: *P. syringae*; *Xanthomonas eleusinae*
  Specific host: Finger millet and pearl millet

Major symptoms
The disease came to knowledge initially from India (Desai et al., 1965) and, thereafter, from Africa (Mudingotto et al., 2002; Adipala, 1980). The symptoms during the diseased condition are as follows:

- During the diseased condition, the first part of the plant that will be susceptible for infection is leaves.

- Infection on millets starts with the appearance of small irregular spots with a characteristic straw-colored center.
- Linear-shaped spots could be observed along the vein on both lower and upper surfaces of leaves.
- During the initial stage, light yellow- or brown-colored spots may be observed, which, with disease progression, turn dark brown in color.
- At a later stage, streaks may be observed on the peduncle.
- The infected plants showed premature wilting

Control

- Before sowing, treatment of seeds with systemic fungicides could prove to be beneficial for preventing the symptoms of disease.
- Use disease-resistant cultivars.

9.3.2 Leaf Stripe

Disease-causing microorganisms: *Pseudomonas avenae; Pseudomonas eleusinae*
  Specific hosts: Finger millet and pearl millet

Major symptoms

- Long, narrow, red-colored stripes on leaves.
- Infected leaf sheath shows brown coloration.
- Straw-colored midrib.
- Light brown discoloration on one side.

9.3.3 Leaf Streak

Disease-causing microorganisms: *Xanthomonas axonopodis pv. Pennamericanum; Xanthomonas axonopodis pv. Coracanae; Pseudomonas avenae*
  Specific hosts: Pearl millet, small millets, foxtail millet, barnyard millet, proso millet

Major symptoms
The disease is most commonly observed in millets and the symptoms are as follows:

- Narrow, interveinal, water-soaked streaks (pale yellow, brownish with red margins)
- Bacterial exudates on lesions
- Broad yellow-colored lesions which, upon disease progression, turns into brown-colored ones
- Premature wilting in plants may be observed if the infection starts at the early stage of growth

Control

- Use of disease-resistant millets cultivar.
- Pretreatment of seeds with hot water.
- Removal of undesired weeds from the field helps to eradicate disease symptoms.
- Supply of balanced nutrients/biofertilizer helps the plant to gain immunity/resistant against the disease.
- Avoid unnecessary addition of water in the field.
- Spraying copper-based specific microcidals may help provide resistant against infection.

9.4 VIRAL DISEASES IN MILLETS

Virus-mediated infections have also been reported in millets, and the onset of disease could be expressed in sporadic forms under favorable environmental conditions.

9.4.1 Ragi Severe Mosaic

Disease causing organism: Sugarcane mosaic virus

Symptoms

- Stunted growth
- Malformed ears
- Production of small-sized seeds
- Less yield
- Chlorosis

In India, during the rainy season, the viral-mediated disease symptoms have been reported in Andhra Pradesh and Karnataka. Due to viral infection in millet crops, the infected plants failed to set seeds in them (Joshi et al., 1966). Viral-induced mosaic symptoms have been clearly observed in young leaves. Plants infected with the disease showed stunted growth with malformed ears. With disease progression, the infected plant produces seeds that are comparatively smaller in size, ultimately resulting in the less crop yield. In addition to stunted growth, infected plants also suffer from chlorophyll loss, resulting in pale yellow appearance. Due to yellow color, the infected plants could be easily distinguished from healthier plants. The infection may occur in millet crops at any stage of their growth.

9.4.2 Ragi Mottle Streak

The first report on Ragi mottle streak virus was also documented from Karnataka (Mariappan et al., 1973). Maramorosch et al. (1977) reported 50%–100% reduction in crop yield due to Ragi mottle streak virus.

Symptoms

- Appearance of dark green color along leaf veins
- Chlorosis
- Stunted growth
- Small ears
- White specks

9.4.3 Ragi Streak

Disease-causing organism: Eleusine strain of maize streak virus

Symptoms

- Streaking
- Yellow color in leaf
- Stunted growth
- Less yield

- Reduction of grain weight
- Pale specks on young leaves
- Chlorotic bands

During the initial stage of viral infection, yellow color is visible in leaves with streaking. Further, as the disease enters the next phase, the plants show stunted growth with chlorosis. Depending on the severity of the disease and age of the plant, significant reduction in yield was observed, as indicated by decreased grain weight by 24%–84% (Nagaraju et al., 1982; Anonymous, 1975). Chlorotic bands/streaks may be observed in the infected plants. High number of tillers with yellow-colored sickly ears bearing shriveled grains is one of the important symptoms of the disease.

9.5 PRE- AND POSTHARVEST MANAGEMENT

Preharvest evaluation helps to ensure the quality of grains being recovered during the harvest process while leaving the unripe crop aside. During the postharvesting time, one of the most important steps is to ensure the quality of crop, especially whether ripening is completed uniformly throughout the field. The different steps in the pre- to postharvesting of millets are explained in the following.

9.5.1 Preparation of Land

Preparation of land for growing millets is an important step. Millets can be grown in different soils. Preparation of land is a necessary step to eliminate undesirable plants or weeds from the farm. If weeds are not properly removed, they will grow with the crop and hinder crop growth. Preparation of land also provides a favorable condition for germination of the seed and good growth of a plant. Drying and plowing land in the summer season kill the weed seeds, insects, and disease-causing organisms by exposing them to heat. The second main benefit of plowing and then irrigating the dry land is that the moisture lasts a long time. After plowing the dry land, the field is watered with available resources. Before the soil becomes lump, plowing is done. For plowing the farm, disc

(a)   (b)

(c)   (d)

Figure 9.3 Agricultural machinery used for land preparation: (a) cultivator, (b) rotavator, (c) disc harrow, and (d) suhaga land leveler.

harrow, cultivator, and rotavator can be used (Figure 9.3). After plowing the land two or three times, land is leveled using suhaga land leveler. Land leveling is necessary to enable the placement of seeds at a certain depth and maintain even depth for all seeds.

9.5.2 Selection of Seeds

After the preparation of land, the second step is sowing the seed. According to the fertility of land, required water resources, and climatic conditions, the seed variety is selected. Hybrid varieties of millets are available, and their yields are high. Requirements for seed selection are shown in Table 9.2. Seed requirement is less when line-sowing method is opted while broadcasting method requires a higher amount of seeds. Foxtail, kodo, proso, and barnyard millets required 8–10 kg/ha seeds when the line method is opted, while 15 kg/ha is needed for the broadcasting method.

Table 9.2 Amount of Seed of Millets Required Per Hectare

Crop	Amount of Seed Per Hectare (kg/ha)
Pearl millet	3
Finger millet	8–10
Foxtail millet	10–15
Kodo millet	10–15
Barnyard millet	10–15
Proso millet	10–15

9.5.3 Seed Treatment

Seed treatment refers to the exposure of seeds to certain physical, chemical, or biological agents. These are not only employed to make the seeds pest- or disease-free but also to provide the possibility of pest and disease control during germination and emergence of young plants and early growth of plants (Forsberg et al., 2003). Seed treatment requires adequate standards and good application methods. These should be agreed upon between the involved parties on a case-by-case basis to ensure producer, applicator, seed, and environmental safety in the most cost-efficient manner. Seed treatment with biopesticides (*Trichoderma harzianum* @ 4 g/kg) or thiram 75% dust @ 3 g/kg seeds will help prevent soil-borne diseases. Seed treatment with 300-mesh sulfur powder @ 4 g/kg seeds controls smut disease. For removing ergot-affected seeds, soaking in 10% salt solution is needed. Seed treatment with metalaxyl (Apron 35 SD) @6 g/kg seeds controls downy mildew. Seeds are treated with *Azospirillum* (600 g) and *Phosphobacterium* to enhance the availability of nitrogen and phosphorus.

9.5.4 Sowing of Seeds

Different methods are used for sowing seeds in the farm. Line spacing between rows and plants varies according to the crop. Normally, three methods are used for sowing the millets: line sowing, broadcasting, and nursery method. Inline methods are used for proper sowing of the seeds. In this line spacing method, spacing between rows and plants is properly maintained. In this method, fewer seeds are required as compared to the broadcasting method. Broadcasting is a common method of seed sowing;

in this method, seeds are spread on the soil, and seeds may or may not be covered with soil. Seeds can be spread manually or through a mechanical spreader. Pearl, proso, kodo, foxtail, and barnyard millets are generally sown by line method or broadcasting method. The seeds should be sown at a depth of 2.5–4.0 cm according to millet type. Finger millet is sown using the nursery method. Nursery method is a very common method for paddy and some vegetables. First, a nursery is prepared; then, the plants are grown to sufficient height and then transferred to an already-prepared field.

9.5.5 Manures and Fertilization

Nutrients are essential for growth and yield of crops. All crops need nutrients to grow, and many nutrients are destroyed during harvesting of the crops. For proper growth of crops, balanced amount of nutrients are required. To recover these nutrients, synthetic fertilizers or manures are used. Farmers can get their farm soil checked to know what nutrients are lacking. Plants take macronutrients from the soil to fulfil their requirement. Mineral nutrients are normally provided to plant in the form of fertilizers. A fertilizer includes minerals like nitrogen, phosphorus, potassium, sulfur, calcium, magnesium, etc. Micronutrients are also essential for proper growth of plants. Micronutrients such as molybdenum, boron, copper, zinc, and iron are also provided to crops in sufficient amounts. Farmers also use organic manures (farmyard manure, compost, green manure, etc.) to improve the nutrient content of soils. Organic manure is the best method to compensate the nutrient content of soil. Organic manure, 5–10 tons, is applied to the farmland before watering and preparation of land. Urea and NPK are used for normal growth of the plants (Figure 9.4).

9.5.6 Irrigation

Millet is a drought-tolerant crop. Compared with other cereal crops, millets require less water. Under prolonged dry spells, irrigation should be applied at critical stages of crop growth, i.e., tillering, flowering, and grain developmental stages, if water is available. Irrigation is done with tubewells and canal water. Depending on the soil type, weather condition, and variety, irrigation is done.

238

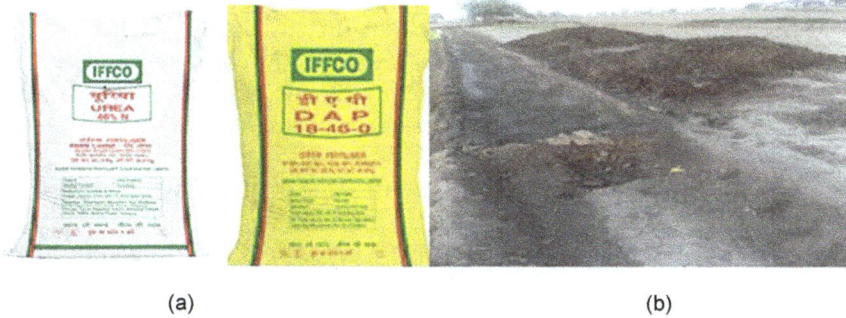

(a)   (b)

Figure 9.4   Types of fertilizers: (a) synthetic fertilizers and (b) manures.

9.5.7 Weed Management

Weed management is essential for the growth of plants; if weed content is high, it will utilize the nutrients that should be utilized by the millet plants. Two times intercultivation with blade hoe at 3 and 5 weeks of germination will help make available sufficient aeration, control weeds, and conserve moisture. The losses due millet include (1) direct yield losses due reduced crop quality; (2) increase cost in harvesting, land preparation, and similar operation; and (3) harboring of insect pests and diseases. Major weeds of millets are *Cynodon dactylon* (Scutch grass), *Brachiaria ramose* (Browntop millet grass), *Digitaria sanguinalis* (hairy crabgrass), *Dactyloctenium aegyptium* (crowfoot grass), *Dinebra retroflexa* (viper grass), *Chloris barbata* (windmill grass), *Eleusine indica* (yard grass), *Echinochoa colona* (deccan grass), *Setaria glauca* (Zipati grass), *Setaria viridis* (green bristlegrass), *Convolvulus aruensis* (bindweed), *Acanthospermum hispidum* (bristly starbur), *Achyranthes aspera* (prickly chaff flower), *Commelina benghalensis* (Benghal dayflower), *Ageratum conyzoides* (goatweed), *Amaranthus palmeri* (carelessweed), *Amaranthus retroflexus* (redroot pigweed), *Boerhaavia diffusa* (punarnava) and *Celosia argentea* (plumed cockscomb) (Mishra et al.,2018).

9.5.8 Disease Management

Diseases are one of the major constraints in millet production. In future, these problems are likely to increase because of climate change. Diseases in millets have significant effects as they result in economic as well as

239

postharvest losses in the grain production sector throughout the world. Detection of disease-causing pathogens in millets is crucial for the maintenance of sustainability in farming sector. Major diseases of millets are grain mold, downy mildew, blast, smut, rust, ergot, leaf spot, brown spot, and, occasionally, viral and bacterial diseases (Das, 2017). Various factors that play vital roles in occurrence of diseases in crops are temperature, rainfall, soil profile, pH, moisture, growth stage of plants, wind, deficiency of nutrients, weak immune system, and chemical-induced damages (Jahromi, 2007; Velásquez et al., 2018). Under the favorable conditions, opportunistic microorganisms play their roles and cause diseases in crops. More than 85% of diseases in crops are caused by fungal strains, followed by bacteria and viruses.

9.5.9 Harvesting

Harvesting of millets could be performed by removal of heads with sharp hand knives/sickles; in some regions, the activity is performed by breaking the stems (Esele, 1989). For a few days, the harvested ears are kept at ambient conditions for further ripening of grains with specific taste. The harvested crop may be kept under sunlight to keep the moisture content balanced and to ensure their natural drying.

9.5.10 Inspection and Crop Certification

Inspection of harvested crop ensures the quality of grains/seeds to be stored. Certification is another important step that helps to determine the specific cultivars so that superior-quality grains could be made available for public/farmers.

9.5.11 Transport

After harvesting, transport of millet crops may be done immediately or after drying the harvested crops in the field. Transport of harvested crops merely depends on the choice of farmers as sometimes they prefer to dry the crop in the field/house or otherwise they pack the harvested crops followed by their transport. The remaining part of plants after harvesting could be used as fuel sources and animal feed (Seetharam et al., 1989).

9.5.12 Drying

Drying is necessary to stop the microbial infections as well as to improve the shelf life of the crop. Drying is also important for the maintenance of moisture of grains at suitable level. Drying of grains could be performed under sunlight or using specific equipment like hot air oven. In Africa, people prefer rain-beaten grains as traditionally this was assumed to improve the palatability and quality (McFarlane et al., 1995).

9.5.13 Cleaning

Cleaning of grains is an important step as it helps in separation of debris, soil particles, and contaminants. Contamination in the main crop may be due to small stones, shriveled seeds, leaves, animal hairs, sticks, dead insects, excreta of animals and birds, and broken and weak seeds. Removal of metal pieces is also necessary as they may create problems during mechanized grinding and milling. Further, cleaning ensures the taste and quality of products based on millet grains. There are two common methods that are adopted for cleaning grains: winnowing and aspiration.

Winnowing

Winnowing is a wind-mediated cleaning method that has been adapted to separate impurities (straw/pest/soil/foreign contaminants) on the basis of weight. In this method, air is allowed to pass through grains to separate unwanted lightweight impurities from heavy grains using winnowing fans.

Aspiration

In this method, a rotary drum is employed to separate the contaminants from grains. The method uses coupling of horizontal scourers and a vibro separator to improve the grain cleaning efficiency.

9.5.14 Packaging

After cleaning grains, the next step is packaging. In this step, grains are packed in pest-free suitable sisal/jute bags of desired capacity

(50, 100 kg, etc.). Millet grains could also be packed in plastic polybags to ease transportation.

9.5.15 Storage and Assessment of Crop Loss

Storage of millet grains is an important aspect in the production system as it facilitates the year-round availability, avoids shortage of food materials, and serves as a seed reservoir. Traditional methods include storage of millets in mud straw bins, reeds, sealed drums, and mud walls (Esele, 1989). The shelf life of stored grains mainly depends on the conditions under which they are being stored. The quality of grains could be maintained using appropriate temperature conditions and humidity in the storage room. Loss to millets crop could be evaluated using different approaches (Krall et al., 1995; Nwanze, 1988). The results may help to evaluate the loss as per differences in infection and yield per hectare. A comparative analysis of year-wise production could also help to understand the disease severity and its impact on millets.

9.6 HOST PLANT RESISTANCE

An effective strategy is important for the determination of germplasm and its use in developing suitable breeding programs (Nagaraja & Das, 2016). Scarcity of information on millet diseases makes the crop more important from the researcher's point of view. A few reports are available on blast disease and resistance to natural infections (Nagaraja et al., 2010; Kumar & Kumar, 2009; Kumar et al., 2006; Mantur et al., 2001). Disease-causing fungal strains have the potential to enter into the plant system through the stomatal opening or by epidermal cell piercing. Madhukeshwara (1990) reported the difference during the earlier stages of infection among cultivars. Further, disease-resistant cultivars possess higher granulation at cytoplasmic levels as compared to susceptible host cultivars (Madhukeshwara et al., 1997). During the onset and progression of disease, the pericarp and endospermic region showed the presence of fungal spores/hyphae, whereas the embryo does not (Pande et al., 1994). Anatomical reports on healthy and infected plants showed significant differences in the thickness of the cuticle layer. High thickness was observed in resistant cultivars as compared to disease-susceptible host (thin layer). Further, size difference and low stomatal frequencies per square millimeter have been observed for resistant cultivars as compared to susceptible hosts (Sanathkumar et al., 2002). Kumar and Kumar (2009) reported the presence of smut (grain and head) and brown spot resistance in barnyard millet (TNAU-92, VL-216, and PRB-402). Further, seven different foxtail millet genotypes (GPUS-27, TNAU-213, TNAU-235, SiA-3088, SiA-3059, SiA-3039, and SiA-3066) showed no evidence of brown spot disease. One of the significant problems in disease management of millets is that poor farmers are growing millets without much knowledge on millet diseases; also, the lack of information on resistant cultivars creates various problems (Gowda et al., 1986). The conventional breeding program is strong enough to fight against disease conditions, so it generates the necessity of molecular approach to achieve targeted resistance in cultivars (Ignacimuthu & Ceasar, 2012). Byregowda et al. (1997) demonstrated that variations in genotypic and phenotypic expressions play an important role while managing diseased conditions. Improvement at the genetic level is necessary to combat the problems of diseases in millets.

9.7 FUTURE PROSPECTS

To prevent the frequent occurrence of diseases in millets, it is necessary to continue the evaluation of virulence, identify the disease-causing organisms, understand their life cycles, identify resistance factors, and develop resistant cultivars. Understanding the relationship between changing environmental conditions and disease occurrence may help to prevent various diseases in millets. During disease, thorough assessment of risk on model plants may help the researchers collect detailed information on crop, disease occurrence, and impact of surrounding environmental factors. These factorial analyses help to design tools for the preparation of disease-resistant cultivars. Detailed information helps breeders to select cultivars especially according to the region/climatic conditions so as to minimize the risk of disease occurrence. Research policies should be prepared in accordance with the location-based identification of disease, identification of resistant/pyramiding genes, and marker-assisted appropriated linkage mapping so as to assist in solving the disease-related issues of millets.

9.8 CONCLUSIONS

To better understand the disease, it is of utmost importance that one should understand the disease-causing mechanism followed by causal organisms and identification of factors that can restrict the growth of the causative agent. Developing molecular strategies to impart resistance to the plant could prove to be an efficient management tool. Disease-controlling biocontrol agents should be popularized at the commercial scale to minimize the loss in millet crops. Millet crops and their diseases have attracted the attention of molecular biotechnologists/agro-industrial sector throughout the world. Use of effective management strategies, considerable progress while understanding the relationship between hosts and pathogens, screening methods, and keen interest in disease-controlling factors could help the agricultural sector see new success. Apart from diseases that have a major impact on millets, those that pose low risk (minor diseases) should also be given importance so that basic information, management protocols, and epidemiology could be systematized.

REFERENCES

Abdou Zayan, S. 2020. Impact of climate change on plant diseases and IPM strategies. Plant diseases - current threats and management trends. http://dx.doi.org/10.5772/intechopen.87055

Adipala, E. 1980. Diseases of finger millet (Eleusine coracana (L.) Gaertn) in Uganda (M.Sc thesis). Makerere University. 186 pp.

Andersen, E., Ali, S., Byamukama, E., Yen, Y. and Nepal, M. 2018. Disease resistance mechanisms in plants. Genes 9: 339.

Anonymous, 1975. Annual report of the virologist. In: AICRP on small millets 197475. Bangalore: University of Agricultural Sciences.

Baltes, N. J., Gil-Humanes, J. and Voytas, D. F. 2017. Genome engineering and agriculture: opportunities and challenges. Gene Editing in Plants: 1–26.

Bhat, R. V., Shetty, H. P. K., Amrut, R. P. and Sudershan, R. V. 1997. A food borne outbreak disease due to the consumption of moldy sorghum and maize containing fumonisin mycotoxins. *Journal of Toxicology – Clinical Toxicology* 35: 249–255.

Bilgrami, K. S. and Choudhary, A. K. 1998. Mycotoxins in pre-harvest contamination of agricultural crops. In: Sinha, K. K. and Bhatnagar, D., editors. *Mycotoxins in agricultural and food safety.* New York: Marcel Dekker. pp. 1–44.

Butler, E. J. 1918. *Fungi and diseases in plants.* Calcutta: Thacker Spinck and Co. p. 547.

Byregowda, M., Shankaregowda, B. T. and Seetharam, A. 1997. Associations of biochemical compounds with blast disease in finger millet. Extended Summary, National Seminar on Small Millets. ICAR and TNAU, Coimbatore, p. 54.

Cunniffe, N. J., Laranjeira, F. F., Neri, F. M., DeSimone, R. E. and Gilligan, C. A. 2014. Cost-effective control of plant disease when epidemiological knowledge is incomplete: modelling bahia bark scaling of citrus. *PLoS Computational Biology* 10: e1003753.

Dang J. K. and Panwar, M. S. 2004. Downy Mildew of Pearl Millet: Present Scenario in India. In: Spencer-Phillips P. and Jeger M., editors. *Advances in Downy Mildew Research-Volume 2. Developments in Plant Pathology,* vol 16. Dordrecht: Springer. pp 165–178. https://doi.org/10.1007/978-1-4020-2658-4_10

Das, I. K. 2017. Millet diseases: current status and their management. *Millets and Sorghum: Biology & Genetic Improvement* 10: 291–322.

De-Carvalho, A. de O., Soares, D. J., do Carmo, M. G. F., da Costa, A. C. T. and Pimentel, C. 2006. Description of the Life-cycle of the Pearl Millet Rust Fungus–*Puccinia substriata* var. penicillariae with a Proposal of Reducing var. indica to a Synonym. *Mycopathologia* 161(5): 331–336.

Desai, S. G., Thirumalachar, M. J. and Patel, M. K. 1965. Bacterial blight disease of *Eleusine caracana* Gaertn. *Indian Phytopathology* 28: 384–386.

Esele, J. P. 1989. *Cropping systems, production technology, pests and diseases of finger millet in Uganda in small millets in global.* Seetharam, A., Riley, K. W. and Harinarayana, G., editors. Ottawa, Canada: IDRC.

Frederiksen, R. A., Castor, L. L. and Rosenow, D. T. 1982. Grain mold, small seed and head blight: the Fusarium connection to head in sorghum. *Proceedings of the 37th Annual Corn and Sorghum Industry Research Conference* 37: 26–36.

Forsberg, G., Kristensen, L., Eibel, P., Titone, P. and Haiti, W. 2003. Sensitivity of cereal seeds to short duration treatment with hot, humid air. *Journal of Plant Disease & Protection* 110(1): 1–16.

Gowda, B. T. S., Seetharam, A., Viswanath, S. and Sannegowda, S. 1986. Incorporating blast resistance to Indian elite finger millet cultivars from African cv. IE 1012. *SABRAO Journal* 18: 119–120.

Herve, M., Albert, C. H. and Bondeau, A. 2016. On the importance of taking into account agricultural practices when defining conservation priorities for regional planning. *Journal for Nature Conservation* 33: 76–84.

Hückelhoven, R. 2007. Cell wall–associated mechanisms of disease resistance and susceptibility. *Annual Review of Phytopathology* 45: 101–127.

ICRISAT (International Crops Research Institute for the Semi-Arid Tropics) 1987. *Proc Internat Pearl Millet Workshop,* 7–11 April 1986, ICRISAT Center, India, Patancheru, p. 36.

245

Ignacimuthu, S. and Ceasar, S. A. 2012. Development of transgenic finger millet (*Eleucine coracana* (L.) Gaertn.) resistant to leaf blast disease. *Journal of Biosciences* 37: 135–147.

Indira, S. and Rana, B. S. 1997. Variation in physical seed characters significant in grain mold resistance of sorghum. In *Proceedings of the International Conference on Genetic improvement on Sorghum and Pearl Millet*, Lubbock, 23–27 September 1996, INTSORMIL and ICRISAT. pp. 652–653.

Jahromi, F. G. 2007. Effect of environmental factors on disease development caused by the Fungal pathogen *Plectosporium alismatis* on the floating-leaf stage of starfruit (*Damasonium minus*), a weed of rice. *Biocontrol Science & Technology* 17(8): 871–877.

Joshi, L. M., Raychaudhuri, S. P., Batra, S. K., Renfro, B. L. and Ghosh, A. 1966. Preliminary investigations on a serious disease of *Eleusine coracana* in the states of Mysore and Andhra Pradesh. *Indian Phytopathology* 19: 324–325.

Jurjevic, Z., Wilson, J. P., Wilson, D. M. and Casper, H. H. 2007. Changes in fungi and mycotoxins in pearl millet under controlled storage conditions. *Mycopathologia* 164: 229–239.

Juroszek, P. and Von-Tiedemann, A. 2011. Potential strategies and future requirements for plant disease management under a changing climate. *Plant Pathology* 60: 100–112.

Kaur, P. and Purewal, S. S. 2019. Biofertilizers and their role in sustainable agriculture. In: Giri, B., Prasad, R., Wu, Q. S. and Varma, A., editors. *Biofertilizers for sustainable agriculture and environment*. Soil Biology. Vol. 55. Cham: Springer. pp. 285–300.

Kaur, P., Purewal, S. S., Sandhu, K. S., Kaur, M. and Salar, R. K. 2018. Millets: a cereal grain with potent antioxidants and health benefits. *Journal of Food Measurement & Characterization* 13: 793–806.

Krall, S., Youm, O. and Kogo, S. A. 1995. Panicle insect pest damage and yield loss in pearl millet. In: Nwanze, K. F. and Youm, O., editors. *Proceeding of an International Consultative Workshop on Panicle Insect Pest of Sorghum and Millet*, ICRISAT Sahellan, Centre, Niamey, Niger, pp. 135–145.

Kumar, K., Gambhir, G., Dass, A., Tripathi, A. K., Singh, A., Jha, A. K., Yadava, P., Choudhary, M. and Rakshit, S. 2020. Genetically modifed crops: current status and future prospects. *Planta* 251: 91.

Kumar, B. and Kumar, J. 2009. Evaluation of small millet genotypes against endemic diseases in mid-western Himalayas. *Indian Phytopathology* 62: 518–521.

Kumar, V. B. S., Kumar, T. B. A., Bhat, S. A. and Nagaraju 2006. Screening of long duration finger millet (*Eleucine coracana* (L.) Gaertn.) genotypes against neck and finger blast caused by *Pyricularia grisea* (Cke.) Sacc. *Journal of Plant Protection and Environment* 3: 136–139.

Madhukeshwara, S. S. 1990. Studied on variability in *Pyricularia grisea* (Cke.) Sacc. With particular reference to virulence (M.Sc. thesis), University of Agricultural Sciences, Bangalore, p. 91.

246

Madhukeshwara, S. S., Viswanath, S. and Mantur, S. G. 1997. Variability in different isolates of *Pyricularia grisea* (Cke) Sacc. with special reference to cytoplasmic granulation. In *Extended Summaries, National Seminar on Small Millets, Current Research Trends and Future Priorities as Food, Feed and in Processing for Value Addition*, 23–24 April 1997. TNAU, Coimbatore, 73 pp.

Mahdi, M. T. 1962. Studies on *Tolyposporium ehrenbergii*: the cause of long smut of sorghum in Egypt. *Proceedings of the International Test Association* 27: 184–191.

Mantur, S. G., Viswanath, S. and Kumar, A. T. B. 2001. Evaluation of finger millet genotypes for resistance to blast. *Indian Phytopathogy* 54: 387.

Maramorosch, K., Govindu, H. C. and Kondo, F. 1977. Rhabdo virus particles associated with mosaic disease of naturally infected *Eleusine coracana* (finger millet) in Karnataka state (Mysore) South India. *Plant Disease Report* 61: 1029–1031.

Mariappan, V., Natarjan, C. and Kandaswamy, T. K. 1973. Ragi streak disease in Tamil Nadu. *Madras Agricultural Journal* 60: 451–453.

McFarlane, J. A., John, A. E. and Marder, R. C. 1995. Storage of sorghum and millets: including drying for storage, with particular reference to tropical areas and the mycotoxin problem. In: Dendy, D. A. V., editor. *Sorghum and millets: chemistry and technology*: St Paul: MNAACC International. pp. 169–183.

Mishra, J. S., Kumar, R., Upadhyay, P. K. and Hans, H. 2018. Weed management in millets. *Indian Farming* 68(11): 77–79.

Mudingotto, P. J., Veena, M. S. and Mortensen, C. N. 2002. First report of bacterial blight caused by *Acidovorax avenae* ssp. avenae associated with finger millet seeds from Uganda. *Plant Pathology* 51: 396.

Nagaraja, A. and Das, I. K. 2016. Disease resistance in pearl millet and small millets. *Biotic Stress Resistance in Millets*: 69–104.

Nagaraja, A. and Mantur, S. G. 2007. Screening of *Eleusine coracana* germplasm for blast resistance. *Journal of Mycopathology Research* 45: 66–68.

Nagaraja, A., Reddy, N. Y. A., Gowda, J. and Reddy, A. B. 2010. Association of plant characters and weather parameters with finger millet blast. *Crop Research* 39: 123–126.

Nagaraju, Viswanath, S., Reddy, H. R. and Lucy Channamma, K. A. 1982. Ragi streak a leaf hopper transmitted virus disease in Karnataka. *Mysore Journal of Agricultural Sciences* 16: 301–305.

Nakayama, H., Nagamine, T. and Hayashi, N. 2005. Genetic variation of blast resistance in foxtail millet (*Setaria italica* (L.) P. Beauv.) and its geographic distribution. *Genetic Resources & Crop Evolution* 52: 863–868.

Navi, S. S., Bandopadyay, R., Reddy, R. K., Thakur, R. P. and Yang, X. B. 2005. Effects of wetness duration and grain development stages on sorghum grain mold infection. *Plant Disease* 89: 872–878.

Nene, Y. L. and Singh, S. D. 1976. A comprehensive review of downy mildew and ergot of pearl millet. In *Proceedings on Consultants' Group Meetings on Downy Mildew and Ergot of Pearl Millet*, 1–3 Oct 1975, ICRISAT Center, India, Patancheru. pp. 15–53.

Niks, R. E. and Rubiales, D. 2002. Potentially durable resistance mechanisms in plants to specialised fungal Pathogens. *Euphytica* 124: 201–216.

Nwanze, K. F. 1988. Assessment of on-farm losses in millets due to insect pests. *International Journal of Tropical Insect Science* 9: 673–677.

Nyvall, R. F. (1989). *Field Crop Diseases Handbook*. Springer, US. doi:10.1007/978-1-4757-5221-2

Pande, S., Mukuru, S. Z., Odhiambo, R. O. and Karunakar, R. I. 1994. Seed-borne infection of *Eleucine coracana* by *Bipolaris nodulosa* and *Pyricularia grisea* in Uganda and Kenya. *Plant Disease* 78: 60–63.

Pennisi, E. 2010. Armed and dangerous. *Science* 327: 804–805.

Prakash, H. S., Nayaka, C. S. and Kini, K. R. 2014. Downy mildew disease of pearl millet and its control. *Future Challenges in Crop Protection against Fungal Pathogens*: 109–129.

Prakash, G., Kumar, A., Sheoran, N., Aggarwal, R., Satyavathi, C. T., Chikara, S. K., Ghosh, A. and Jain, R. K. 2019. First Draft Genome Sequence of a Pearl Millet Blast Pathogen, Magnaporthe grisea Strain PMg_Dl, Obtained Using PacBio Single-Molecule Real-Time and Illumina NextSeq 500 Sequencing. *Microbiology Resource Announcements* 16:e01499-18.

Purdy, L. H. and Kendrick, E. L. 1963. Influence of environmental factors on the development of wheat bunt in the Pacific Northwest. IV. Effect of soil temperature and moisture on infection by soil spores. *Phytopathology* 53: 416–418.

Rai, K. N. and Thakur, R. P. 1995. Ergot reaction of pearl millet hybrids affected by fertility restoration and genetic resistance of parental lines. *Euphytica* 83:225–231.

Salar, R. K. and Purewal, S. S. 2016. Improvement of DNA damage protection and antioxidant activity of biotransformed pearl millet (*Pennisetum glaucum*) cultivar PUSA-415 using *Aspergillus oryzae* MTCC 3107. *Biocatalysis & Agricultural Biotechnology* 8: 221–227.

Salar, R. K. and Purewal, S. S. 2017. Phenolic content, antioxidant potential and DNA damage protection of pearl millet (*Pennisetum glaucum*) cultivars of North Indian region. *Journal of Food Measurement & Characterization* 11: 126–133.

Sanathkumar, V. B., Anilkumr, T. B. and Nagaraju 2002. Anatomical defence mechanism in finger millet leaves against blast caused by *Pyricularia grisea*. In *Abstr. Proc. IPS Symp. Plant Disease Scenario in Southern India*, 19–21 December. 15 p.

Satya, P. and Sarkar, D. 2018. Plant biotechnology and crop improvement. *Biotechnology for Sustainable Agriculture*: 93–140.

Savary, S., Ficke, A., Aubertot, J. N. and Hollier, C. 2012. Crop losses due to diseases and their implications for global food production losses and food security. *Food Security* 4: 519–537.

Saveetha, K., Sankaralingam, A., Pant, R. and Ramanathan, A. 2007. Etiology and transmission of mottle streak disease of finger millet (*Eleusine coracana* Gaertn.). *Archives of Phytopathology and Plant Protection* 40(1): 53–60.

Seetharam, A., Riley, K. W. and Harinarayana, G. 1989. *Small millets in global agriculture*. Ottawa, Canada: IDRC. p. 392.

Sharma, R., Upadhyaya, H. D., Manjunatha, S. V., Rai, K. N., Gupta, S. K. and Thakur, R. P. 2013. Pathogenic variation in the pearl millet blast pathogen magnaporthe griseaand identification of resistance to diverse pathotypes. *Plant Disease* 97(2):189–195.

Shcherbakova, L. A. 2007. Advanced methods of plant pathogen diagnostics. *Comprehensive & Molecular Phytopathology*: 75–116.

Siroha, A. K., Sandhu, K. S. and Kaur, M. 2016. Physicochemical, functional and antioxidant properties of flour from pearl millet varieties grown in India. *Journal of Food Measurement & Characterization* 10: 311–318.

Somani, R. B. and Indira, I. 1999. Effect of grain mold on grain weight in sorghum. *Journal of Mycology & Plant Pathology* 29: 22–24.

Thakur, R. P. and King, S. B. 1988. Registration of six smut resistant germplasms of pearl millet. *Crop Science* 28: 382–383.

Van Esse, H. P., Reuber, L. and Vander Does, D. 2019. GM approaches to improve disease resistance in crops. *New Phytologist* 225: 70–86.

Velásquez, A. C., Castroverde, C. D. M. and He, S. Y. 2018. Plant–pathogen warfare under changing climate conditions. *Current Biology* 28: R619–R634.

Verma, R. S. 2013. Genetically modified plants: public and scientific perceptions. *ISRN Biotechnology* 2013: 1–11.

Williams, R. J. 1984. Downy mildew of tropical cereals. *Advances in Plant Pathology* 3: 1–103.

249

INDEX

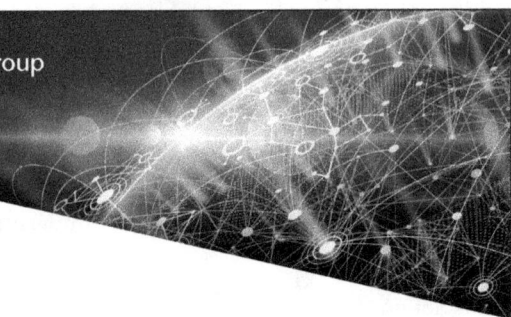

For Product Safety Concerns and Information please contact our EU
representative GPSR@taylorandfrancis.com
Taylor & Francis Verlag GmbH, Kaufingerstraße 24, 80331 München, Germany

www.ingramcontent.com/pod-product-compliance
Lightning Source LLC
Chambersburg PA
CBHW060350220326
41598CB00023B/2864